RF Circuits For 5G Applications

Scrivener Publishing
100 Cummings Center, Suite 541J
Beverly, MA 01915-6106

Publishers at Scrivener
Martin Scrivener (martin@scrivenerpublishing.com)
Phillip Carmical (pcarmical@scrivenerpublishing.com)

RF Circuits For 5G Applications

Designing with mmWave Circuitry

Edited by
Sangeeta Singh
Rajeev Kumar Arya
B.C. Sahana
and
Ajay Kumar Vyas

Scrivener Publishing

WILEY

This edition first published 2023 by John Wiley & Sons, Inc., 111 River Street, Hoboken, NJ 07030, USA and Scrivener Publishing LLC, 100 Cummings Center, Suite 541J, Beverly, MA 01915, USA

For more information about Scrivener publications please visit www.scrivenerpublishing.com.

Wiley Global Headquarters
111 River Street, Hoboken, NJ 07030, USA

For details of our global editorial offices, customer services, and more information about Wiley products visit us at www.wiley.com.

Library of Congress Cataloging-in-Publication Data

ISBN 978-1-119-79192-8

Cover image: Pixabay.Com
Cover design by Russell Richardson

Set in size of 11pt and Minion Pro by Manila Typesetting Company, Makati, Philippines

Printed in the USA

10 9 8 7 6 5 4 3 2 1

Contents

Preface

The wireless communication sector is experiencing exponential expansion, particularly in mobile data and the fifth generation (5G) mobile network. This creates fresh market possibilities for designing this industry's integrated circuits (ICs). Due to its many intrinsic benefits, including its ability to mass produce digital ICs at a low cost and its scalable feature size, which allows the integration of base-band DSPs and low-power mmWave analogue circuitry, CMOS technology has so far fulfilled this function well. This makes CMOS a good contender for constructing 5G circuitry. Scaling CMOS circuits, however, is no longer advantageous. Due to less efficient metal interconnect close to the substrate, this scaling trend is now showing a bottleneck in terms of the maximum achievable figure of merit (fmax). The maximum achievable quality factor of these passive on-chip devices and VDD scaling also aggravate the design trade-off, affecting device linearity, integration, and noise immunity. Additionally, connection and parasitic losses become a significant factor due to the BEOL metal stack's scaling down near the substrate at these scaled dimensions.

Furthermore, various FOM factors conflict with the design requirements for mmWave and RF design. In order to solve these competing trade-off needs for CMOS devices in the "Dark Silicon Era" for 5G band communication circuitry, the research community has invested a lot of time and energy. FinFETs are currently outperforming traditional CMOS device architectures at 22nm scaled nodes. Based on this discovery, researchers have investigated several FinFET-based circuits for 5G circuitry.

Chapter 1 discuss the characterization, specialization, and requirements of the 5G network at an introductory level. The frequencies range of (30-300 GHz) and higher than this is the operating range of 5G wireless networks. The 5G network has three basic situations: massive machine-type interchanges. Improved versatile broadband and super solid low idleness interchanges comprise the majority of the necessity of a 5G organization. It represents a confounded arrangement of prerequisites to circuit and foundation originators. Millimeter waves are used to increase the additional capacity for higher throughput, and basically, the mmWave is known as the fastest of 5G. In a MIMO system, multiple continuous data is transferred through a technique known as beamforming. 5G networks connect virtually through everyone, including machines.

Chapter 2 discussed the RF circuit for a 5G system. This chapter provides the path from 5G circuits from requirements to system models and analysis. The communication system model provides in detail RF system analysis with a system-level RF performance model. Link budget defines the overall performance calculation of different predictable and non-predictable factors. The last multiuser multibeam analysis is explained with concluding remarks.

Chapter 3 covers a nonlinear demonstration of AlGaAs–GaAs HBT's characterization in DC. Minimal signal and noise are mentioned. Conjointly some sets of equations squares measure thought-about to require into consideration in noise equations. This general method can derive parameters from other microwave devices, such as MESFETs and high negatron efficiency semiconductors (HEMTs). This chapter will see associate in nursing as a method for extracting interconnected parameters that are incontestable for a manufactory HBT from wherever we are going to show some wonderful results. The primary technique for obtaining the model parameter values of analogous circuit models is parameter extraction by fitting the model responses to measurements. Parameter extraction has traditionally relied on large-signal measurements and DC parameters. The derived versions are appropriate for DC, small, and large signal research.

In chapter 4, we will learn about the practical applications of the multi-harmonic load-pull simulation technique for designing nonlinear microwave circuits. A structured project using multi-harmonic load-pull simulation delves the results of every harmonic ending on the circuit staging, also finding flawless load at every harmonic. Systems performance can be improved notably if we use this systematic design procedure. This approach comes up with an efficient means of nonlinear microwave circuit design. Its superiorities are embellished by the design of a frequency doubler and two power amplifiers.

Chapter 5 reviews common RF terminology and concepts, followed by a discussion of technology-related difficulties, focusing on passives. This chapter is focused on covering the basic microwave RF designing concepts and various existing technologies. The subsequent sections of this chapter cover the various microwave RF figures of merits, its characterization techniques, and RF CMOS co-designing approaches.

Chapter 6 discussed filter fundamentals and primary design methodologies to attain the large gain bandwidth and additional methodology of the

Bode-Fano (B-F) limit. The first section deals with the fundamental RLC band-pass filter. In this section, the quality factor of the filter and the noise are concisely reminded to set the basis of resonant circuits. These circuits are generally amplifiers and oscillators for mmWave application. The following section presents fourth-order filters that improve gain-bandwidth over the classical RLC tank. The main focus of the next section is transformer-based resonators. The parasitic interwinding capacitance consequence has been discussed that provides instinct on the operation of the circuit. Further, this conversation is stretched to attain impedance transformation to understand the power dividers & combiners.

Chapter 7 provides the basic architecture and characteristics of the LNAs for 5G networks. The designs of LNAs have been carefully considered for a variety of 5-G applications. Its architecture method varies depending on whether the frequency band in question is narrowband or wideband. Various topologies for achieving better-optimized circuit efficiency are discussed here. LNAs are usually found at the receiver's front end, absorbing the antenna's input signal and amplifying it with minimal noise. The most important consideration factor for the design of an LNA is the gain and noise. Different noise figures for single-phase, and multistage amplifiers are illustrated in this chapter.

Chapter 8 is based on the mixer designs. Several mixer specifications were considered. In addition, several examples were considered for the illustration. With a detailed analysis of works from several research groups, diode-mixers and several features of transistor mixers, including their various forms, were discussed in detail. Mixers are an essential component of 5G communication architecture. Their incorporation into 5G communication architecture and subsequent implementation alongside other components will be critical for the future of 5G.

Chapter 9 discussed the VCO for the transceiver, the RF circuit, and the types of VCOs. This chapter has presented the trade-offs in the design of VCO. Low-power LC-VCOs have been discussed as well as a design strategy for RF VCOs. It is feasible to establish whether an oscillator covering the frequency range for the target application is viable using the theory offered and some basic information about the existing IC technology.

Chapter 10 describes the basic design and characteristics of the RF Power Amplifier (PA) are discussed in depth. RF Power Amplifier designs are considered for various 5G and higher frequency applications. Here various approaches are used to achieve and optimize the performance of the Power Amplifier. Both the designing Bipolar Power Amplifier and CMOS Power Amplifier are discussed here. This chapter covers the designing parameters for RF Power Amplifiers, the classification of Power Amplifiers, and different designing approaches to design the Bipolar PA and CMOS PA for the application in 5G. Linearization principles are also discussed at the end of the chapter.

Chapter 11 discussed oscillator design. The main building block of an oscillator is the amplifier and a frequency-selective network in positive feedback. A significant requirement for oscillators used in RF applications should consist of proper amplitude-controlled circuitry, low phase error, and low power consumption. This chapter presents the important aspects of the design of RC and LC oscillator circuits.

Chapter 12 discussed designing millimeter wave-based CMOS PAs (Power Amplifiers), which work in the broadband spectrum. Here, the fundamentals of power amplifier designing and associated difficulties involved in millimeter wave operations are discussed. Many methodologies are given for understanding the cancellation of wideband distortion and load impedance, though the 2-way power combining is allowed to enhance the delivered output power. A few complexities of AB-type power amplifier operations, viz. efficacy at power back-off, main reasons behind amplitude modulation-phase modulation distortions, and techniques to have liner PAs are mentioned. Finally, the last section confers the design and layout of a 29 to 57GHz (65% bandwidth) amplitude modulation-phase modulation compensated AB class PA designed for 5G phased arrays.

Chapter 13 presented the scaling effects on RF performance of Fin devices, including the parasitic and noise components. This chapter also focused on the impact of self-heating and temporal process variability on the electrical performance of Fin devices.

Chapter 14 describes the significance of a power amplifier (PA) as maintaining a promising place for transmitting information in a vast space. Many circuits of the power amplifier in recent times utilize GaN (Gallium nitride) device, which plays a crucial part in highly efficient PA design. Several classes of power amplifiers (D, E, F) theoretically have high efficiency (100%). The linearity measure has become an important factor in the characterization of PA. Day by day, the modulation methods are becoming advanced, and it eases the process of linearizing an amplifier setup. This chapter presented an update and an overview of power amplifier (PA) linearization. In this chapter, we have discussed basic measures of power amplifier nonlinearities. The effect of those nonlinearities and various types of linearization techniques are discussed. Analog pre-distortion (APD) finds better space in the communication system field. One composite structure of a hybrid EFJ power amplifier with an APD linearizer block is developed to verify the improved linearization, and simulation results are compared with the state-of-the-art linearization schemes.

Chapter 15 discussed one such antenna selection scheme for the system with the combined Massive MIMO and mmWave technologies. This chapter also describes the Massive MIMO operations supported by millimeter-wave (mmWave) technologies which operate in the frequency band of 30-300 GHz. At the same time, a discrete lens array mechanism is adopted to reduce the system's energy consumption.

Chapter 16 represented the control circuit for reduced power dissipation region due to shadow effects in solar photovoltaic affecting the performance at the module level. Furthermore, the decreased energy in any particular module creates a mismatch at the string level, affecting the overall system performance. Techniques such as MPPT (Maximum Power Point Tracking), Bypass Diode, Bridge Linked (BL), and Total Cross Tied (TCT) are used to mitigate the shading effects up to some extent. Still, none of them are much effective in tackling this issue properly. The experimental results cause 27.78% and 55.56% more energy generation compared with conventional module architecture with 72 cells by activating one and two bypass diodes, respectively.

Sangeeta Singh
Rajeev Kumar Arya
B.C. Sahana
Ajay Kumar Vyas
December 2022

Part I

5G COMMUNICATION

1

Needs and Challenges of the 5th Generation Communication Network

Anamika Raj[1*], Gaurav Kumar[2] and Sangeeta Singh[1]

[1]*Microelectronics & VLSI Design Lab National Institute of Technology, Patna, India*
[2]*Muzaffarpur Institute of Technology, Muzaffarpur, Bihar, India*

Abstract

This chapter discusses characterization, specialization and the requirements of the 5G network at preliminary level. The frequencies range of (30–300 GHz) and higher than this is the operating range of 5G wireless networks. Fundamentally, there are three basic situations of 5G network, which are huge machine-type interchanges. Improved versatile broadband, super solid low idleness interchanges, comprises the majority of necessity of a 5G organization. It represents a confounded arrangement of prerequisites to circuit and foundation originators. Millimeter waves are used to increase the additional capacity for higher throughput and basically, mmWave is known as fastest of 5G. In MIMO system multiple continuous data is transferred through a technique known as the beamforming. 5G networks connect virtually through everyone together including machines also. 5G networks will give higher download rates of up to 10 gigabits each second (Gbit/s). Consequently, it is anticipated that 5G networks have in excess of 1700 million supporters all throughout the planet by 2025, according to the reports of GSM affiliation.

***Keywords*:** 5G, mmWave, OFDM, CDMA, LTE

1.1 Introduction

In 2008, NASA Machine-to-Machine Intelligence (M2Mi) Corp to develop IoT and M2M technology, which supported 5G technology, South Korea

**Corresponding author*: anamikar.pg19.ece@nitp.ac.in

Sangeeta Singh, Rajeev Kumar Arya, B.C. Sahana and Ajay Kumar Vyas (eds.) RF Circuits For 5G Applications: Designing with mmWave Circuitry, (3–18) © 2023 Scrivener Publishing LLC

also started working in the field of 5G and instituted many research and development programs. We are now transforming ourselves from automation era to intelligent machines era in which the decision making capability of devices will be enhanced to the next level, for this purpose a large number of data is required which should be transmitted at a very high speed and this enhanced superior technology that enables us to connect devices with a very high speed will bring transformation everywhere and it will affect our day to day life, from business operations to smart home, unmanned vehicle to driverless car, from banking to healthcare. It also opens the door for telemedicine, remote surgeries, and even sometime remote monitoring that will save life up to a great extent.

In 5G a new technology known as nr or new radio, it was developed by 3rd generation partnership project (3GPP) for 5G (fifth generation) mobile networks. It is actually a new radio access technology or RAT was designed and developed for the standardization of air interface of 5G mobile networks. It is very contrasting with the earlier existing mobile technologies. It has high-speed range of internet service. The majority of domains are still under the research phase and we can expect a huge number of use case is yet to come. 5G offers a very high speed and it supports large numbers of devices that can digitize many industrial aspects. It can work in high just as low-recurrence ranges. Talking about data rate 5G offers speed of 10Gbps in the downlink with ultra-low latency of 1 ms. 150Mbps is the lower average speed of the 5G.

Millimeter waves are used by several network operators for additional capacity, and for higher throughput and basically mmWave is known as fastest of 5G. It has a more restricted reach than microwaves, so the cells are confined to gauge. Millimeter wave is more unobtrusive than the tremendous getting wires used in cell associations. Sometimes it is only in the range of a few centimeters long.

In 4G, the concept of Multiple-input Multiple-output (MIMO) was introduced and at each cell it uses 32 to 128 small antennas in very beginning of 2016. In the proper configuration and frequencies, the performance can be expanded by four to multiple times. Different piece of information are sent simultaneously utilizing a technique known as beamforming, the PC at base station will compute the best course for the radio waves to arrive at every one of the remote gadget in the blink of an eye and will arrange a various receiving wires to cooperate as clusters to make light emissions waves to arrive at the each gadget. 5G networks require high transfer speed, inclusion, accessibility with low idleness so exceptionally high requests of quicker correspondence can be obliged. Some significant necessities of a 5G organization, which are acquiring acknowledgment of the business, are enlisted below:

i. Bandwidth: 1–10Gbps
ii. Latency: 1 millisecond
iii. Network energy usage: 90% decrease from 4G
iv. number of associated gadgets: 1–100 times of 4G
v. Battery life expectancy: Approximately 10 years (for low power handsets)
vi. Coverage: 100%
vii. Availability: 99.99%

1.1.1 What is 5G and Do We Need 5G?

5G stands for 5th Generation Mobile Network. It comes after 1G, 2G, 3G, and 4G networks. The invention of 5G network is not by the single person, but there are many companies within the mobile ecosystem that bring 5G to life. Many Companies has played a crucial role in the invention of the numerous fundamental innovations that make up the 5G, the next wireless standard network as Figure 1.1 shows the expansion of the 5G technology. 5G Network empowers us another sort of organization through which we can interface essentially everybody together including machines, gadgets and so forth. 5G wireless technology means to deliver followings:

- Higher multi-Gbps top information speeds
- It has massive network capacity
- 5G has ultra-low latency

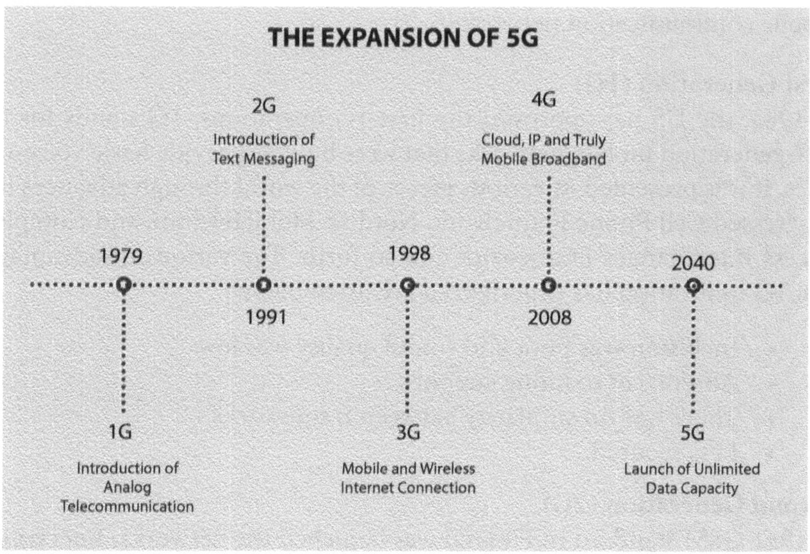

Figure 1.1 The expansion of 5G.

- More reliable
- Increases availability
- Higher performance
- 5G improves efficiency and empower new user experiences
- Uses wider bandwidth technologies

The basic working principle of 5G is based on OFDM (Orthogonal frequency-division multiplexing). 5G OFDM deals with comparative compact framework organization norms. Notwithstanding, the new 5G NR air interface can update OFDM to pass on significantly more genuine degree of flexibility and versatility. This technique gives more 5G induction to more people and in wide scope of usage cases.

5G brings widens the ranges of bandwidths by enlarging the spectrum of resources, mainly from 3 GHz used in 4G to 100 GHz and beyond this. 5G can work in both lower groups just as mmWave that will bring outrageous limit, low idleness and multi-Gbps throughput. It is planned not exclusively to convey quicker speed however better portable broadband administrations contrasted with 4G LTE, it very well may be likewise ventured into new help zones, for example, strategic interchanges and interface the IoT [1–5].

1.1.2 A Brief History of Gs

Every successive generation is abbreviated with "G" in the wireless standard, which introduce the confounding advances in data-carrying capacity and reduces the latency as increasing the G increases the facilities in the mobile communication network [6, 7].

First Generation (1G)

In 1983, the US has approved the first 1G operations. 1G stands for the first-generation mobile networks that were built to provide basic voice services. It was presented in various pieces of the world through advances like Progressed Cell Phone Framework, Nordisk MobilTelefoni, and Complete Access Interchanges Framework and so forth. The various disadvantages that 1G innovation has experienced are listed below:

- Inclusion was poor and sound quality was low.
- Absence of roaming support.
- There was no similarity between frameworks.
- Less secured.

Second Generation (2G)

In 1991 GSM standard in Finland was launched the networks. Encryption feature that was absent in the first Generation was now introduced For the first time, calls was encrypted and the quality of digital voice was significantly improved and was much clearer than earlier generations with less background crackling and people sent text messages (SMS), picture, and multimedia messages (MMS), on phones. The second generation of mobile networks introduces two new technologies Time Division Multiple Access and Code Division introduced Multiple Access. The analog history of first generation gave way to the digital future to 2G. This led to mass-adoption among consumers and businesses [8].

2G's trade speeds were from the start around 9.6 Kbit/s, dashed to place assets into new system, for instance, versatile cell towers. Close to the finish of this age, paces of 40 Kbit/s were feasible and after that EDGE affiliations offered speeds of up to 500 Kbit/s. Despite by and large having sluggish rates, 2G modified the business and changed the world forever.

Third Generation (3G)

In 2001 NTT DoCoMo with an aim to standardize the protocols of network 3G was launched. Using CDMA technology, there are two key methods for 3G. Universal Mobile Telecommunications Systems (UMTS) being the first track while the second one was CDMA2000. UMTS was acclimated with 5migrating the GSM associations to 3G and CDMA2000 was the 3G advancement for IS-95 and D-AMPS. UMTS uses Wideband CDMA for the passage part and offers paces of up to 2 Mbps [9]. customers could get to data from any side of the world as the 'data packages' that drive web accessibility which made worldwide wandering organizations curiously a veritable chance.3G's expanded information move multiple times quicker than 2G additionally prompted the ascent of new administrations, for example, video real time, video conferencing and voice over IP like Skype. Despite all of these advantages, the cost of cellular infrastructure is very high in 3G.

Fourth Generation (4G)

4G was first introduced in 2009 as the Drawn-out Development (LTE) 4G standard. It was as needs be introduced ridiculous and made amazing video electronic for countless customers. 4G offers speedy convenient web access which works with HD chronicles, gaming organizations, etc. Long haul Advancement is the 4G relocation way for key 3G innovations including General Portable Media transmission Framework (UMTS) and CDMA2000. Innovation like Overall Interoperability for Microwave Access give a 4G overhaul way however LTE has been the essential innovation utilized worldwide for 4G. LTE is significantly more productive than the previous 3G advances, and it diminishes the inactivity in the information move. After the dispatch of LTE, LTE Progressed (LTE-A) and LTE Star were presented. LTE can uphold up to 300 Mbps speed in the downlink, while LTE-Master and LTE-A which can uphold greatest velocities of up to 3Gbps and 1Gbps individually [10]. While 4G is current norm all throughout the planet however in certain locales has confronted network inconsistency.

Fifth Generation (5G)

5G stands for fifth generation technology in telecommunications for broadband cellular networks. It is more competent, unified air interface. It is intended to stretch out ability to empower future, engage new models and convey new administrations. With superior reliability, high speeds, ultra-low latency, 5G expands the mobile ecosystem into new era. 5G impact on every industry, making precision agriculture, safer transportation, digitized logistics, remote healthcare and many more.

5G are cell organizations, in which administration region is isolated into the little geological regions called cells. All gadgets of 5G remote in a cell are associated with phone and Internet network in neighborhood receiving wire in the cell by radio waves. The principle benefit of this new organization is that they will give higher download speeds, up to 10 gigabits each second (Gbit/s). It has expanded transfer speed, thus, it is normal that organizations won't only serve cell phones, yet additionally it is utilized as broad web access suppliers for PCs and work area rivalling accessible ISPs as satellite web, and will make new applications in and machine to machine zones and web of things (IoT). It is expected that 5G associations have more than 1700 million endorsers worldwide by 2025, according to the GSM association [10–13].

1.2 mmWave Spectrum, Challenges, and Opportunities

Millimeter band is also known *as* Millimeter wave (MMWave), is the band of spectrum with wavelength between 10 millimeters and 1 millimeter and frequency between (30 GHz) to (300 GHz). International Telecommunication Union (ITU) called this as the extremely high frequency (EHF) band. Regardless of the benefit of contiguous spectrum in mmWave, cell sort of correspondence network with mmWave innovation has been considered as trying. Fundamentally, because of ominous channel qualities of mmWave range, which decrease an assistance inclusion and existing broadband mobile network. The challenges basically consist of:

- Large path loss
- Impact of atmospheric absorption O_2, CO_2

- Rain and fog attenuation
- Mobility support has been limited.

These difficulties can likewise be sorted into various gatherings:

- Spectrum aspects
- Propagation aspects
- Energy efficiency aspects
- Cost aspects.

Due to previously mentioned reasons of mmWave potential arrangement situations innovation have been believed to be restricted reach highlight point correspondence in a (LOS) view with low versatility. Nonetheless, ongoing outcomes show that the mmWave correspondence is viable technology for the outdoor cellular communication.

Diverse examination chances of mmWave correspondence in the up and coming age of portable broadband organizations. Various parts of heterogeneous organizations just as multi-radio wire handset innovations are examined as:

- Heterogeneous Networks
- Advanced Multi-Antenna Transceivers

Performance improvement can be gotten by conveying base stations into a nearness of terminals. Coming about are great channel conditions among recipients and transmitters. Thus, a decreased transmission forces can be utilized to diminish obstruction among coinciding networks. It tends normal that co-channel-interference among a few mmWave HetNet layers becomes decreased empowering mmWave little layers to be super thick.

Heterogeneous Networks

Heterogeneous organizations are perceived as standard change in the traditional cellular network characterizing to improve network limit just as administration inclusion territory. As a rule, HetNet have distinctive kind of organization hubs or network nodes outfitted with various handling abilities communicate power spending plans, and backing diverse radio access innovations. In this way, HetNets can be considered as multi-various leveled network with a few overlaying layers. It comprises of mmWave network hubs for access and gadget to-gadget (D2D) correspondence, and backhaul [14].

- Network Densification: Performance improvement can be gotten by conveying base stations into a nearness of terminals. Coming about great channel conditions among recipients and transmitters. Thus, a decreased transmission forces can be utilized to diminish obstruction among coinciding networks. It tends normal that co-channel-interference among a few mmWave HetNet layers becomes decreased empowering mmWave little layers to be super thick.
- Backhauling: It refers to the side of the network that communicates with global Internet. To enable low latency and high data mobile services under 5G networks, backhaul offer a local help by empowering high-limit transfer of data between core-network and access point. Fiber and copper are considered as cost proficient data transmission solutions for locations. For locations without wired network, the structure of wired transmission may have a methodology as far as expenses, i.e., CAPEX and OPEX. In this way, there is a need to have low latency, super high limit, and adaptable just as cost proficient remote backhaul transport answers for 5G mobile networks
- Co-existence of the mmWave with HetNet Layers: To ensure concurrence of mmWave with other HetNet layer, effective between working between conceivably non mmWave, and a few HetNet layers must be empowered. Thus there is a need of plan conjunction techniques for the mmWave communication to be essential for the in general 5G network [15].
- **Advanced Multi-Antenna Transceivers:** To decrease cost, power utilizations, and execution intricacy in mmWave, novel handset models are expected to acknowledge in the multi-gigabit information rates in the down to earth execution. As should be obvious, the vast majority of sign preparing is performed at baseband in discrete-time space before digital to-analog conversion or analog to-digital transformation. However, power utilizations in computerized spatial space preparing engineering are high. The purpose for is that ceaseless transmission transfer speeds related with high inspecting rates in DAC and ADC. Furthermore, each communicates and gets receiving wire cluster components having DAC/ADC unit driving expenses be high. This engineering control both the plentifulness and sign's stage.

The engineering is more adaptable and can give be upheld to information multiplexing and Eigen beamforming [16–19].

1.3 Framework Level Requirements for mmWave Wireless Links

Demands of large number of 5G wireless network are very difficult. It requires another sort of structure for channel. 5G Channel models network incorporate more boundaries or parameters path loss, shadow blurring or fading and beamforming are three proliferation impacts that portray the radio climate of wireless channels.

Path loss: This addresses attenuation with expanding in the division distance between the transmitters (BS) and receivers (UE).

$$PL_1[dB] = 10\log_{10}\frac{P_T}{P_R} \tag{1.1}$$

Where PR is characterized as an element of distance and frequency from Friss Formula we can obtained as given in Equation (1.2).

$$P_R(d,\lambda) = P_T G_T G_R \left(\frac{\lambda}{4\pi d_{3D}}\right)^2 \tag{1.2}$$

- Shadow fading effect: Changes in underlying, topographical and natural highlights like territory abnormalities, vehicles building type, and other infrastructural snags influence signal propagation between Base Station and the Client Hardware, which shows deviation from the normal path loss values. This is called as shadow blurring (SF), which when joined with the path loss is the explanation of compelling path loss. SF is demonstrated as a distance-subordinate log-typical distribution N (μ, σ), with zero mean (μ) and standard deviation (σ) that relies upon the transporter recurrence. Narrowband flat fading channel, limited scope fading can be given as in the piece of channel impulse response

$$h(t, \tau) = V_1 + g(t, \tau) \tag{1.3}$$

Where, V characterizes a complex and destined segment, which characterized the view path or line of sight way between the transmitter and recipient. On the off chance that various radio waves are received wide-sense fixed uncorrelated scattering (t, τ) is a complex boggling zero-mean Gaussian random variable with the Rayleigh distribution

Beamforming: It empowers MIMO frameworks; MIMO and beamforming are at times utilized conversely. Beamforming is utilized in mMIMO. When all is said in done, beamforming is essentially used in multiple antennae to control the bearing of a wave-front by fittingly weighting the stage and size of individual antenna signals in the array of numerous antennas, which is, the very signal that is sent from multiple antenna and that has adequate room between them. In the given area, the receiver will get same copies of signal. On the area of the receiver, the signals are in different phases, averaging each other out, or helpfully summarize in the various copies are in same phase. For mmWave remote frameworks or wireless systems, and leading impediments and numerous dielectric of this present reality will cause way misfortune in contrast with free space. To improve the current path loss (PL) models, a nitty gritty thought of certifiable boundaries must be finished.

Digital beamforming: Pre-coding or baseband beamforming are other terms for digital beamforming. The antenna elements' phase and amplitude have been pre-customized to improve the cell limit. With the help of persuasive asset management (frequency/time), this allows synchronized transmission of information or data for several clients. At the same time, a similar configuration of antenna elements can be used to shape multiple bars (one for each client).

Analog beamforming: It changes the gain and amplitude of the antenna array, which permits fractional compensation for high path loss at mmWave frequencies. Be that as it may, simple beamforming just permits the bar for a bunch of antenna elements.

Hybrid beamforming: It is a mix of both analog and digital beamforming plans one strategy is to utilize analog and digital beamforming for coarse and fine, separately.

1.4 Circuit Aspects

Being a circuit designer, one should have knowledge about the circuit performances, power losses of the circuit and simply we can say we should have data sheet of the circuit. That will tell us the higher and lower operating range of that specific circuit.

While designing any circuit we should have a proper mathematical expression, design aspects which will be validated and verified with the help of simulations at very preliminary stage. From the mathematical and theoretical aspects, we will got the different technical key aspects or parameters like operating frequencies, operating voltages, current flowing, heat dissipation, ideal and practical temperature specification, gain, losses, etc., basically we are preparing the sheets of technical specification and after simulation, testing verification , quality check we then finally prepare the best technical sheets All the process is very rigorous and it is done in a feedback loop until the desired and optimum is obtained.

As we realize that 5G networks work in frequencies scope of (30–300 GHz), which is fairly large as compared with 4G and 3G. The accessible channel transmission capacity of 4G is 20 MHz 5G requirements a huge transfer speed to communicate extremely high date traffic, high increase power enhancers which commands the plan of rapid. Such plan prerequisites push inconsistent message plan as far as possible. Printed circuit loads up should oblige both simultaneously. The channel data transfer capacity for the 5G networks lies in the reach from 100 MHz to 400 MHz.

The transfer speed per channel is a significant perspective that requires better approaches to plan PCB materials as it assumes a fundamental part in the presentation of a framework. Ongoing coordinated circuits can oversee both high frequencies and high information traffic. Be that as it may, PCB needs consideration regarding guarantee brilliant execution for 5G organization. It requires a high velocity and high-recurrence inconsistent message framework. PCB and the material segment have fundamental job in forestalling signal misfortunes and guaranteeing signal trustworthiness. Fashioners have rules for plan and format for improving the working conditions. Contradicting message frameworks comprises both simple and computerized and it is inclined to Electromagnetic obstruction. PCBs ought to be intended to forestall EMI between various areas of the board, and configuration ought to have electromagnetic similarity. This requires lower dielectric steady (~3.0) and these new materials will ready to substitute polytetrafluoroethylene for 5G remote frequencies.

Thermal management is helpful to limit the varieties in the yield power amplifiers and preparing units. It gets basic at high frequencies as thermal coefficient and thermal conductivity of the dielectric constant changes quickly with temperature. Anyway, dielectric steady of protectors adheres to reverse proportionality law with the temperature. A thermal runway causes the deficiency of dielectric execution and an increment the force utilization of the framework.

To get dispersed heat away from dynamic devices quickly, PCBs should be based on a thermally conductive substrate. It will reduce the variety in the yield of handling units and force intensifiers at greater speeds and frequencies by balancing the current in circuits. The warm administration structure will help to reduce scattering. It connects variations in parasitic capacitances' RC time constants, changes in interconnect spread speed, changes stretch computerized heartbeats, and prompts signal reflections to transmission lines in extreme circumstances [20].

1.5 Outline of the Book

Almost all modern cellular mobile communications take place in the sub-3 GHz band, which has now become too congested to handle future needs for increased mobile data traffic. As a result, employing the 28 GHz and 39 GHz millimeter wave (mmW) frequency bands, a new paradigm for constructing next-generation cellular communication systems (i.e. 5G) has been investigated. The RF/microwave industry's near obsession with the coming Fifth Generation (5G) of wireless communications technology dominates the flavor of this book, especially in the area of integrated circuits, this is especially true (ICs). This book gives readers a paranoiac view and in depth analysis of the analog IC designing approaches for realizing the 5G communication circuit. Due to intrinsic benefits such as cost-effective mass manufacture of digital ICs and scalable feature size, which allows the integration of low-power mmWave analogue circuitry with base-band digital signal processing units, CMOS technology has traditionally fulfilled this function well.

This book starts with the introduction of 5G circuitry and its designing challenges and what are the various designing approaches that are followed for the 5G circuit. Then a brief introduction of 5G circuits from requirements to system models and various modelling approaches are covered. It also includes HBT high-frequency modeling and integrated parameter extraction approaches. Post this it depicts the nonlinear microwave circuit design using multi-harmonic load-pull simulation techniques and microwave RF designing concepts and technology are discussed. Then, the in depth design of various RF circuit components such as low noise amplifiers (LNA), mixer, RF LC VCOs, RF power amplifiers, and RF oscillators designing is covered. We've gone through several design considerations to keep in mind when creating these circuits, as well as the negative scaling implications. We've added real-world RFIC design challenges, case studies based on experimental results, and clearly defined design guidelines for 5G communication ICs to improve the readability of this book. As a result, CMOS is a strong candidate in the construction of 5G circuits. Furthermore, the design requirements of mmWave and RF design are incompatible in terms of key figures-of-merit parameters. As a result, the research community has worked hard to resolve these opposing trade-off needs for CMOS devices in the "Dark Silicon Era" for 5G band communication circuitry. FinFETs have recently outperformed conventional CMOS device architectures at 22nm scaled nodes. Hence, this book covers the FinFET based circuitry designing approaches for the 5G communication network as well.

Acknowledgement

The authors are grateful to National Institute of Technology, Patna for supporting this chapter writing in terms of computational requirements.

References

1. Everything you need to know about 5G, Accessed: Apr. 22, 2020. [Online]. Available: https://www.qualcomm.com/invention/5g/what-is-5g.
2. Staff, S.D., 5G technology promises faster connections, lower latency, Accessed: Apr. 22, 2020. [rdquo [Online]. Available: https://www.sdxcentral.com/5g/definitions/5g-technology/.
3. ITU-R The International Telecommunication Union, *IMT Vision–Framework and Overall Objectives of the Future Development of IMT for 2020 and Beyond*, ITU-R The International Telecommunication Union Accessed: Apr. 23, 2020, Available online: https://www.itu.int/dms_pubrec/itu-r/rec/m/R-REC-M.2083-0-201509-I!!PDF-E.pdf.
4. Kavanagh, S., 5G.co.uk, What is enhanced mobile broadband (eMBB), 5G.co.uk, Accessed: Apr. 27, 2020. [Online]. Available: https://5g.co.uk/guides/what-is-enhanced-mobile-broadband-embb/.
5. ITU-T L.1310–study on methods and metrics to evaluate energy efficiency for future 5G systems, Accessed: Apr. 23, 2020. Available online: http://handle.itu.int/11.1002/1000/13476.
6. 3GPP TS 22.261 V17.1.0, 3GPP, *3rd Generation Partnership Project; Technical Specification Group Services and System Aspects; Service Requirements for the 5G System*, 3GPP, Valbonne, France, Accessed: Apr. 23, 2020, Available online: http://www.3gpp.org/ftp//Specs/archive/22_series/22.261/22261-h10.zip.
7. Kozma, D., Varga, P., Soos, G., Supporting digital production, product lifecycle and supply chain management in industry 4.0 by the arrowhead framework-

a survey, in: *IEEE International Conference on Industrial Informatics (INDIN)*, July 2019, pp. 126–131, 2019.
8. Lie, D.Y.C., Mayeda, J.C., Li, Y., Lopez, J., A review of 5G power amplifier design at cm-wave and mmWave frequencies. *Wirel. Commun. Mob. Comput.*, 2018, 1–16, Jul. 2018.
9. Niida, Y., Kamada, Y., Ohki, T., Ozaki, S., Makiyama, K., Minoura, Y., Okamoto, N., Sato, M., Joshin, K., Watanabe, K., 3.6 W/mm high power density W-band InAlGaN/GaN HEMT MMIC power amplifier, in: *PAWR 2016-Proceedings of the 2016 IEEE Topical Conference on Power Amplifiers for Wireless and Radio Applications*, pp. 24–26, 2016.
10. Lie, D.Y.C., Tsay, J., Hall, T., Nukala, T., Lopez, J., High-efficiency silicon RF power amplifier design-current status and future outlook, in: *RFIT 2016-2016 IEEE International Symposium on Radio-Frequency Integration Technology*, pp. 1–4, 2016.
11. Lee, L.B., Bogale, T.E., Wang, X., Chapter 9–mmWave communication enabling techniques for 5G wireless systems: A link level perspective, a paradigm for 5G, in: *mmWave Massive MIMO: A Paradigm for 5G*, pp. 195–225, 2017.
12. Lombardi, R., ETSI, *Microwave and Millimetre-Wave for 5G Transport, White Paper*, First Edition, ETSI, CEDEX, France, February 2018, Available online: https://www.etsi.org/images/files/ETSIWhitePapers/etsi_wp25_mwt_and_5g_FINAL.pdf [Accessed: 23-Apr-2020].
13. 5G requirements: Wireless technology needs » electronics notes, Accessed: Apr. 28, 2020. [Online]. Available: https://www.electronics-notes.com/articles/connectivity/5g-mobile-wireless-cellular/requirements.php.
14. Nguyen, T., Small cell networks and the evolution of 5G (Part 1), Accessed: Apr. 21, 2020. [Online]. Available: https://www.qorvo.com/design-hub/blog/small-cell-networks-and-the-evolution-of-5g.
15. Sun, S., Rappaport, T.S., Shafi, M., Tang, P., Zhang, J., Smith, P.J., Propagation models and performance evaluation for 5G millimeter-wave bands. *IEEE Trans. Veh. Technol.*, 67, 9, 8422–8439, Sep. 2018.
16. Sun, S., Rappaport, T.S., Rangan, S., Thomas, T.A., Ghosh, A., Kovacs, I.Z., Rodriguez, I., Koymen, O., Partyka, A., Jarvelainen, J., Propagation path loss models for 5G urban micro-and macro-cellular scenarios, in: *IEEE Vehicular Technology Conference*, July 2016, 2016.
17. Rupasinghe, N., Kakishima, Y., Guvenc, I., *System-Level Performance of mmWave Cellular Networks for Urban Micro Environments*, Aug. 2017, arXiv:1708.03963v1.
18. Passoja, M., 5G NR: Massive MIMO and beamforming–what does it mean and how can I measure it in the field?, Accessed: Apr. 27, 2020. [Online]. Available: https://www.rcrwireless.com/20180912/5g/5g-nr-massive-mimo-and-beamforming-what-does-it-mean-and-how-can-i-measure-it-in-the-field.

19. Nordrum, A. and Clark, K., 5G bytes: Beamforming explained. *IEEE Spectr.*, Accessed: Apr. 25, 2020. [Online]. Available: https://spectrum.ieee.org/video/telecom/wireless/5g-bytes-beamforming-explained.
20. PCB materials and design requirements for 5G systems, Accessed: Apr. 23, 2020. [rdquo [Online]. Available: https://resources.pcb.cadence.com/blog/2019-pcb-materials-and-design-requirements-for-5g-systems.

2

5G Circuits from Requirements to System Models and Analysis

Vipin Sharma[1*], Rachit Patel[1] and Krishna Pandey[2]

[1]*Department of Electronics and Communication Engineering, ABES Institute of Technology, Ghaziabad, India*
[2]*Department of Electronics and Communication Engineering, National Institute of Technology, Patna, Bihar, India*

Abstract
In present days, in keep view of demand of high data rate, and enabling technologies, i.e., IoT application, drone application, military applications, Artificial Intelligence, there is very much needed requirement of deployment of 5G system in the country. In the chapter RF requirement for 5G systems with Radio spectrum and standardization is discussed. Communication system model provides in detail RF system analysis with system level RF performance model. Link budget defines the overall performance calculation of different predictable and non-predictable factors. In the last multiuser multibeam analysis is explained with concluding remarks. This chapter provides the path from 5G circuits from requirements to System Models and Analysis.

Keywords: 5G communication, MIMO (Multiple input multiple output), OFDM (Orthogonal frequency division multiplexing), system scaling, link budget

2.1 RF Requirements Governed by 5G System Targets

In the world, G. Marconi gives the concept of electromagnetic wave transmission in the era of 1890, although it takes almost 79 years to grow the technology for practical applications and commercial availability. After 1969, the technologies rapidly evolved with introduction of Antenna

**Corresponding author*: vippi.vipin@gmail.com

Sangeeta Singh, Rajeev Kumar Arya, B.C. Sahana and Ajay Kumar Vyas (eds.) RF Circuits For 5G Applications: Designing with mmWave Circuitry, (19–30) © 2023 Scrivener Publishing LLC

design, microwave devices, microelectronics and integrated circuits. Since 1980, the mobile communication grows rapidly with advancement in communication technologies. In present time, even if you talk about the Moore's law, which defines the increase in the density on a single chip, becomes fail. Failure of Moore's law shows the advancement in the manufacturing process, technologies and compactness. The first generation of mobile communication use the analog communication methods, i.e., FDMA (frequency division multiple access) which ensure and supports the voice signals. CDMA (code division multiple access) and TDMA (time division multiple access) are introduced in the 2G generation to enhance its capabilities. 2G generation supports for the voice messages as well as text messages with larger coverage area in comparison to the 1G. Advancement in CDMA introduces the 3G third generation of mobile telecommunication with support of data services. 4G generation evolves with adoption of MIMO and OFDM technologies to enhance the data rate faster than 3G mobile communication. Due to the larger demand of higher data rates, large number of frequency band is needed, and in 4G mobile communication scenario all frequency bands actually lies below than 4 GHz which is susceptible to the environmental conditions. In present's day, the demand of IoT (Internet of things), IoB (Internet of Breath), and IIoT (Industry Internet of Things) applications increases rapidly which actually relies on the wireless communication rather than wired channels. 4G generation somehow unable to meet the demand of higher data rate in case of wireless/mobile communication. That's why the 5G fifth generation replace the prevail technologies with enhancement of higher data rate and frequency bands. 5G mobile communication provides the new paradigm to IoT (Internet of things), IoB (Internet of Breath), and IIoT (Industry Internet of Things) applications. In many countries the deployment of 5G is already done till 2019. In India still, we are looking for the deployment and commercial usage of 5G generation communication. 5G incorporates energy and spectral efficiency with the data rate in Gbps. In 5G millimetre-wave communication higher frequency ranges (6–300 GHz) are unused. To achieve the large number of enabling networks are required with massive MIMO, Ultra-dense networking and full-duplex technologies.

2.2 Radio Spectrum and Standardization

3GPP releases define 5G as the new evolution to the next generation networks; in Release 17, for implementation of 5G NR with NSA deployment and initial specifications. In release 15 in 2018, it gives the SA deployment

Table 2.1 5G NR frequency band specifications.

Frequency band name	Range	Bandwidth (Max)	Subcarrier distance	Area of application
FR1	Below 2GHz	50/100 MHz	15 KHz	In-door coverage
	2GHz-6 GHz	50/100/200 MHz	15/30/60 KHz	To maintain balance between capacity and coverage
FR2	Above 6GHz	200/400 MHz	60/120/240/ 480 KHz	To provide high data rate

dependent on 5GC core networks. In release 16, it extent the prevail technologies with improving the connectivity and communication concern for different enabling technological applications. 5G NR operates in the two frequency bands namely as FR1 and FR2. The detailed range, bandwidth, subcarrier distance, and area of application are mentioned in Table. 2.1.

FR1 operates in two frequency ranges below 2 GHz and 2 GHz to 6 GHz. FR1 (below 2GHZ) is best suited for wide or depth indoor applications and FR1 (2GHz to 6GHz) is utilized for application or areas where the optimized balance is required between the capacity and coverage. The user application requires high data rate than FR2 (above 6GHz) can be used [1].

2.3 System Scalability

Millimetre wave networks and technology provides a prominent solution for spectrum utilization in the 5G communication technologies. But it has also several scalability challenges mentioned as

- ➢ Beam Training
- ➢ Impact of Dynamics
- ➢ Access point sėlection and Handover in Dense networks

Beam training: It is a method which ensures the alignment of antenna beams at the both ends (transmitter and receiver ends). In the 5G networks, high directional are required to minimize the propagation impacts at mmWave frequencies. Due to this beam training is required which decides the best direction for transmitter and receiver. Beam training plays a vital role in mmWave system with providing the facilitation to grab and adaption according to the environmental changes.

Impact of Dynamics: When the communication environment is continuously changing, which actually always occurs in the mobile communication scenario, the beam training cannot perform well because of its threshold limitations. In this situation, the impact of dynamics is a major challenge. However, with the use of location trackers like GPS system, LIDAR, one can easily predict the environmental movements. In recent days, several technologies like machine learning and artificial intelligence evolves. Machine learning technologies and Artificial Intelligence can also be used with mmWave system to find out the probable solution for high dynamicity problem.

Access point selection and Handover in Dense networks: In mmWave networks, when the device density is large and handover is often happening, then fast optimal beam training of each access point is required which may have led to several dysfunctionalities like loss of system capacity or loss of coverage area of access points. For solving this issues, we may require the prior information of networks or scalable information to minimize the overburden on beam training and can deal with environmental dynamics [2].

2.4 Communication System Model for RF System Analysis

In mmWave communication, orthogonal MIMO path is considered for single directed beam at a time. In the analysis of the system practically most common method used the linearity analysis concept although non-linearity of the circuits also produces their effect and ultimately incorporates with the results. Eigen mode analysis (SVD singular value decomposition) is very famous method for the analysis of MIMO system. For the spatial phenomena most of the time we considered the AWGN channel, although for analysis of the practical cases, and taking account of the fading and environmental effects, we can have considered fading channel models. Raleigh fading and Rician fading channel are most famous models which describe the real environmental situations. Rician fading channel may use when there is dense number of users/devices. When we use the phased arrays and directive antenna, then statistical propagation models with map based approach is more useful than past prevailing methods. SNR, throughput and BER most common parameters for the analysis, although when we are dealing with OFDM, several other parameters play a crucial role in the analysis, i.e., coding rate, Maximum achievable data rate etc.

For OFDM data, the Maximum achievable data rate can be defined as

$$Rmax = rc * Nsc * fscs * \frac{\log(M)}{log2} * \frac{tofdm}{tcp + tofdm} \tag{2.1}$$

Whereas, r_c, N_{sc}, f_{scs}, t_{ofdm}, M, and t_{cp} are coding rate, number of subcarrier, subcarrier spacing, ofdm symbol length, modulation order, and cyclic prefix length respectively. With the help equation number 1, we can find out the Maximum achievable data rate of any coding scheme OFDMA modulated variable.

2.5 System-Level RF Performance Model

In the last decade, the demand of Internet is rapidly growing, size of internet users is in cell is densely populated so the new network topology become emerged that deployed the hyper-dense cell into a small cell. Instead of large cell/hyper dense cell, the small cells are preferred by the RF designer due the constraint of theoretical capacity of carrier spectrum. The other aspect of using the micro-cell is to enhance the frequency reuse capability and enlarge the user accommodation capacity as well. The network topology that is deployed by using the micro cell structure is identified as UDN (ultra-dense network) [3].

In 5G UDNs, the massive MIMO antenna arrays accomplished with mmWave has provided almost 1000-time user accommodation capacity as compared to 4th Generation wireless communication, to fulfill the demand of internet capacity. Apart from the user accomodation capacity, networks should maintain the other performance criteria to ensure the QoS [4]. Analytical modelling could be explored by computing the channel parameter for enhancing the system performance and to improved network design [5, 6].

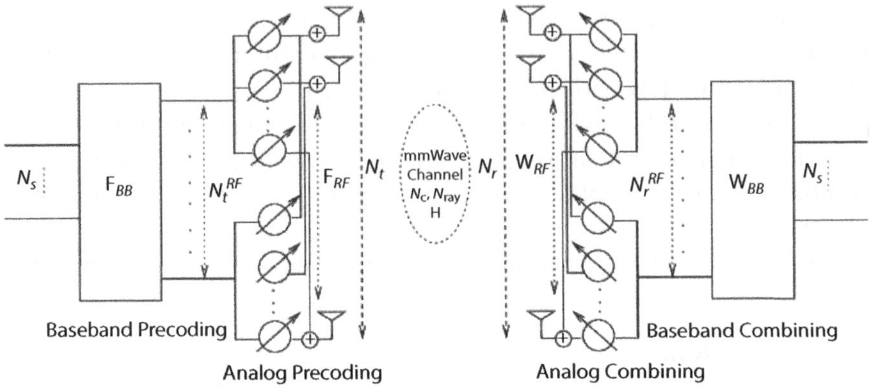

Figure 2.1 Hybrid beamforming architecture.

2.5.1 Transmitter, Receiver, Antenna Array and Transceiver Architectures for RF and Hybrid Beamforming

Hybrid beamforming is a promising solution to enhance the channel capacity by utilizing mmWave accomplished with MIMO antenna array. Earlier, a narrowband channel is utilized to deploy the hybrid (analog & digital) beamforming. However, Analog beamforming is preferred over in broadband system. The biggest advantage to utilizing the hybrid beamforming is to utilize the OFDM Sub-Carrier. Single-user MIMO (SU-MIMO) system is utilized in hybrid beamforming layout deployment at Transmitter and receiver side. To avoid the hardware restriction to deploy the fully-digital beamforming, a noble low dimension for digital and high dimension for analog beamformer approach is utilized in hybrid beamformer [7].

Figure 2.1 shows the transceiver design relying both on a transmit precoder and on analog beamforming. In this design, before the streams are fed to the transmit antennas, they are precoded in the baseband using a linear transmit precoder and then the pre-coded streams are fed to the RF beamformer, where the streams are phase shifted accordingly and finally fed to the transmit antennas [8].

2.6 Radio Propagation and Link Budget

2.6.1 Radio Propagation Model

Radio propagation model is always an important aspect especially when high speed transmission is being involved. In 5th Generation wireless mobile communication system where the technology become deliver the Quality of Services by ensuring high transmission rate, large channel capacity, efficient power saving scheme and ultra-high mobility. 5G is deal with a set of new services as designated with 3GPP release 16 for indoor and outdoor communication. The service requirements to fulfill the objective of 5th Generation communication system need to focus of various performance criteria including latency, security (encryption), spectrum allocation, infrastructure cost, traffic handling capacity, availability of resources, reliability, cast services, channel adaptation ability heterogeneous network, efficient energy/resource allocation are some key points.

In 3GPP there is a lot of space for offering the variety of services in indoor and outdoor environment so in 5th Generation greater flexibility is needed in propagation model. The indoor and outdoor channel modeling and their characterization are very vital property for describing the system performance. The RF propagation model for 5G wireless communication system that more or less depreciate the transmitted signal from the receiver (ground-space propagation model focused on obstacles penetration, reflection, refraction, scattering, diffraction, and absorption by atmospheric gases, moisture). To develop the robust propagation channel model for 5G systems, it is desirable that all the performance measurement criteria should be evaluated well, corresponding to the indoor and outdoor channel model. The linear and nonlinear losses must accountable with link budgeting calculation along with signal power indicator, with accountable of co-channel interference with antenna characterization, along with multipath propagation (random variation in signal strength). In general, it is understanding that only statistic based models are not sufficient to achieve the efficient communicating real-time situation model references will also helpful to obtain appropriate propagation model.

2.6.2 Link Budgeting

The Link budgeting calculation, is a crucial parameter for RF network designer, where all the linear (predictable) and nonlinear (non-predictable) losses must be accountable in prior. Using a Figure 2.2; it can be observing that how link budgeting is implemented [9]. The link budgeting is the prior estimation of linear and nonlinear losses so the end user gets the sufficient signal strength (RxSL). The received signal energy level is compared the desirable signal strength (receiver sensitivity) to identify weather channel link is usable or not. If the received signal strength is passed (desirable), it is said to be reception is good in quality. Link budgeting calculation can be explore using the mention formula.

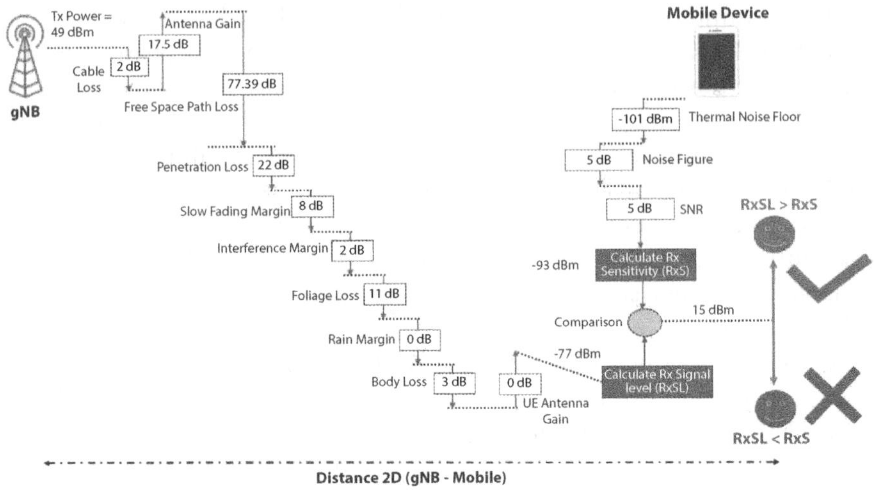

Figure 2.2 Link budgeting parameters.

Received Signal Level at receiver (dBm) = Transmit power by BTS/ gNodeB (dBm) – 10*$\log_{10}$ (sub-carrier power) + Tx antenna gain (dBi) – Path loss (dB) – penetration loss (dB) – foliage loss (dB) – body block loss of Mobile station (dB) – interference margin (dB) – rain/ice margin (dB)– fading margin (dB) + Mobile station Receiver antenna gain (dB) (2).

In the above figure example, if the received signal strength is 77 dBm and the desirable receiver signal threshold is -93 than the status of signal is said to be "Pass". To obtain the value of path loss, the propagation model is adopted as per the 3GPP 36.873.

2.7 Multiuser Multibeam Analysis

In last decade, it has been noted that the boom of wireless communication offers a verity of applications, and the connected developing interest for higher rate transmission. Specifically, 5G frameworks, which are communicated ordinarily, can be likewise utilized for broadband intelligent, and accordingly unicast, transmissions. The fifth era of wireless communication focused on digital video broadcasting (DVB) [10], and its advancement, affirmed in 2014 with the name of DVB-S2X [2], address enlightening models in this sense. Frequency allocation is a challenging task for the variety of service offered by the wireless framework.

Multibeam revolution is a decent answer for the utilization of spectrum. It comprises to frame at the wireless by a multitude narrow beam rather than a solitary broader beam. Multibeam utilize the narrow frequency beam by splitting a broader frequency band which prompts a reuse of the spectrum utilization and a major, capacity of the system could be increased. Consequently, every sub-carrier which describes a beam is a typical channel divided among the user of the zone covered by this beam. Therefore, by adapting a TDMA or CDMA, the users in similar zone or cell can utilize the same frequency band. Apart from the advantage of resource sharing using multiple access technique, a significant interference caused by spectrum sharing has affected the performance of entire system. The performance of system could be improved using managing the beam allocation techniques, implemented at the reception end. We have observed a DVB-S2(X) framework [10, 11], where frequency reuse capability is handled. In that particular scenario the ordinary SUD experiences a severe co-channel interference when it is operated at edge of effected area of beam. The uplink channel (i.e., the link between earth station (gateway) to the wireless) of a multibeam wireless system for multimedia services is shown in Figure 2.3.

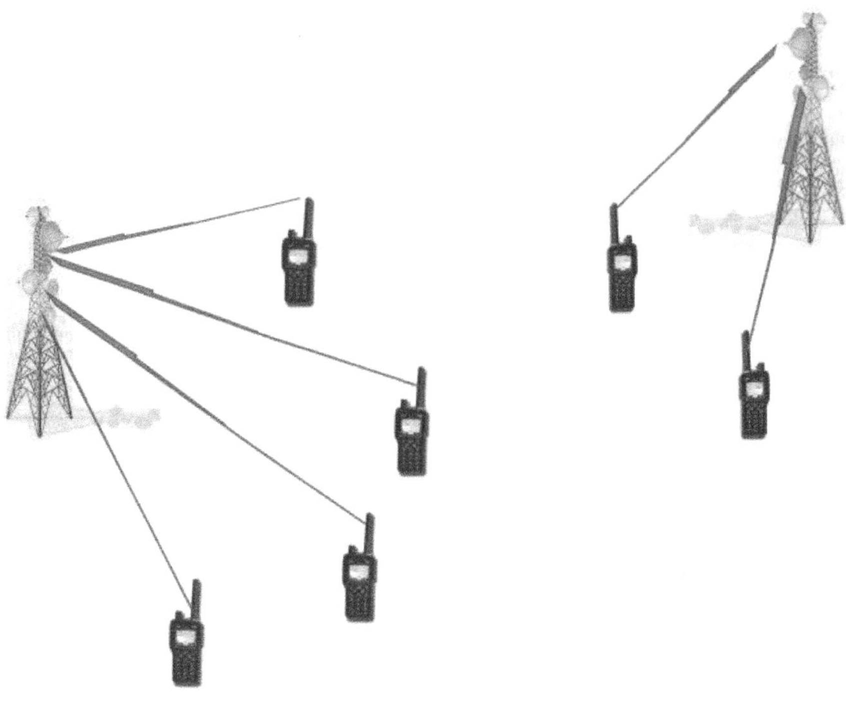

Figure 2.3 Multibeam enabled 5G wireless system.

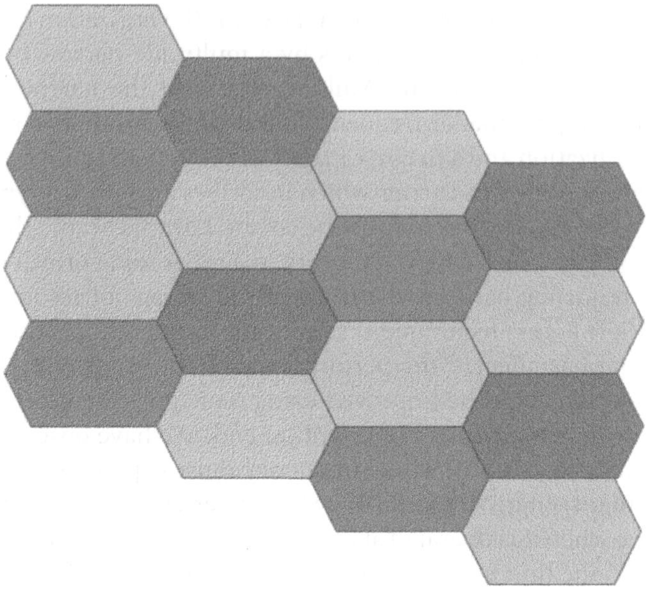

Figure 2.4 Frequency reuse scheme.

In multibeam aspect, the entire covered area is separated into a cell small which is enabled by beams to reuse the frequency in order to enhance the capacity of user's thus the entire spectral efficiency is improved. We are considering, a 4-color cells as shown in Figure 2.4, which shows how multibeam is being efficiently used by ignoring the effect of co-channel interference.

Let us consider a downlink as in Figure 2.3, the beam signals $S_b(t)$, b = 1,.........;N, N is carrier spectrum generated by wireless system separated by a multibeam. Thus the wireless is enable to serve N users by N multibeam poised by wireless transmitters. As wireless operates on multiple carrier signal per transmitter modality so the nonlinear losses could be ignored [10]. The Co-channel interference can also be ignored by increasing the size of the beam cluster.

2.8 Conclusion

5th Generation mobile communication will able to provide the 1000-time traffic demand of current scenario, so as the traffic growth has increased world widely in last few years 5G will definitely ensure QoS with high transmission data rate with ultra-low latency experience. In 5th generation

Radio frequency requirements 3GPP release 16, a variety of carrier spectrum is being designated for indoor and outdoor communication that delivers the high volume data handling capacity, ultra-high mobility. As the variety of spectrum, 5G service are highly scalable. In Radio frequency architecture the linear losses has been covered in link budgeting analysis while nonlinear losses could be mitigated using the adaptive modulation and coding scheme, where appropriate modulation scheme along with channel encoding has been triggered therefore robust communication has been achieved. The MIMO antenna array is being helpful to mitigate the diversity effect and multipath losses.

5G will ensure greater usability of frequency reuse concept by enforcing and multibeam concept where entire covered area is separated into a cell small which is enabled by beams to reuse the frequency in order to enhance the capacity of user's thus the entire spectral efficiency is improved. 5G will also capable to smartly optimize the capacity and multimedia service and it will also helpful to improve energy and cost efficiency. In all we have to recognize the vision of 5G, "information a finger away, everything in touch."

Acknowledgement

The authors are thankful to Dept. of Electronics and Communication Engineering, National Institute of Technology, Patna for providing resources to write this chapter.

References

1. Dogra, A., Jha, R.K., Jain, S., A survey on beyond 5G network with the advent of 6G: Architecture and emerging technologies. *IEEE Access*, 9, 67512–67547, 2021.
2. Fiandrino, C., Assasa, H., Casari, P., Widmer, J., Scaling millimeter-wave networks to dense deployments and dynamic environments. *Proc. IEEE*, 107, 4, 732–745, April 2019.
3. Vahid, S., Tafazolli, R., Filo, M., Small cells for 5G mobile networks, in: *Fundamentals of 5G Mobile Networks*, J. Rodriguez (Ed.), pp. 63–104, John Wiley & Sons Ltd., UK, 2015.
4. Li, Q.C., Niu, H., Papathanassiou, A.T., Wu, G., 5G network capacity: Key elements and technologies. *IEEE Veh. Technol. Mag.*, 9, 1, 71–78, 2014.
5. Ai, B., Guan, K., Li, G., Mumtaz, S., mmWave massive MIMO channel modeling, in: *mmWave massive MIMO: A Paradigm for 5G*, S. Mumtaz,

J. Rodriguez, L. Dai (Eds.), pp. 169–194, Academic Press Cambridge, Massachusetts (US), 2016.

6. Xiao, M. *et al.*, Millimeter wave communications for future mobile networks. *IEEE J. Sel. Areas, Commun.*, PP, 99, 1–1, 2017.
7. Zhang, X., Molisch, A.F., Kung, S.-Y., Variable-phase-shift-based RF-baseband codesign for MIMO antenna selection. *IEEE Trans. Signal Process.*, 53, 11, 4091–4103, 2005.
8. Satyanarayana, K., El-Hajjar, M., Hanzo, L., *Millimeter Wave Hybrid Beamforming with DFT-MUB Aided Precoder Codebook Design*, 2017.
9. Lamali, R. (Solution Architect at Ericsson), 5G network RF planning–Link Budget Basics, Online article (LTE, New Radio, Tech. Fundas), 2019, http://www.techplayon.com/.
10. *ETSI EN 302 307-1 Digital Video Broadcasting (DVB), Second Generation Framing Structure, Channel Coding and Modulation Systems for Broadcasting, Interactive Services, News Gathering and Other Broadband Wireless Applications, Part I: DVB-S2*, ETSI (France), 2020, http://www.etsi.org.
11. *ETSI EN 302 307-2 Digital Video Broadcasting (DVB), Second Generation Framing Structure, Channel Coding and Modulation Systems for Broadcasting, Interactive Services, News Gathering and Other Broadband Wireless Applications, Part II: S2-Extensions (DVB-S2X)*, ETSI (France), 2020, http://www.etsi.org.

3

Millimetre-Wave Beam-Space MIMO System for 5G Applications

G. Indumathi[1*], J. Roscia Jeya Shiney[1†] and Shashi Kant Dargar[2‡]

[1]Department of ECE, Mepco Schlenk Engineering College, Sivakasi, Tamilnadu, India
[2]Department of ECE, Kalasalingam Academy of Research and Education, Anand Nagar, Krishnankoil, Virudhunagar Dist., Tamilnadu, India

Abstract

The primary requirements are high capacity, increased efficiency in the fifth generation, and the forthcoming 6G wireless communication systems. The recent developments of multiple-input, multiple-output (MIMO) systems support high data rates with more accommodation of users within the system. After the evolution of Massive MIMO systems, hundreds of antennas are employed to implement at the base station to provide huge bandwidth capability. At the same time, these large numbers of antennas circuits consume more energy, and the respective hardware costs are also high. Massive MIMO operations can be effectively supported by millimetre-wave (mmWave) technologies which operate in the frequency band of 30-300 GHz. the antenna size at the mmWave frequency range results in tiny antennas. So the combination of MIMO and mmWave technologies can be considered to support effective data rates of the current wireless communication systems. Also, using this combination, 3D beamforming can be obtained but with reasonable loss due to the attenuation of the signals. To cope with this problem, an antenna selection scheme reduces the number of active radio frequency chains with increased energy efficiency and thus reduces the cost of implementing the base stations. This chapter discusses one such antenna selection scheme for the system with the combined Massive MIMO and mmWave technologies. At the same time, a discrete lens array mechanism is adopted to reduce the system's energy consumption.

Corresponding author: gindhu@mepcoeng.ac.in
†*Corresponding author*: rosciashiney@yahoo.co.in
‡*Corresponding author*: drshashikant.dargar@ieee.org

Sangeeta Singh, Rajeev Kumar Arya, B.C. Sahana and Ajay Kumar Vyas (eds.) RF Circuits For 5G Applications: Designing with mmWave Circuitry, (31–52) © 2023 Scrivener Publishing LLC

Keywords: Massive MIMO, mmWave, beamforming, average sum rate, beam allocation, 5G

3.1 Introduction

Massive multiple-input-multiple-output (MIMO) communications, where many antennas are equipped in the cellular base stations to focus radiated energy towards smaller regions of space. So that it potentially allows some orders of upgrading in spectral and energy efficiency. Recently, the frequency range of 30–300 Gigahertz, where millimetre wave technologies operate, supports massive dimensional MIMO operations [1]. These technologies can provide spectrum reuse at a very short distance also. A combined MIMO and mmWave technologies have been considered an essential technique for beyond 5G and even 6G wireless communications, which can support a significant increase in data rates. According to Rappaport *et al.* 2013, broadband cellular communication can achieve a bandwidth of about 1 GHz at a 38 GHz band of frequency. If millimeter waves are coupled with massive MIMO technology, 3D beam-forming can be quickly done, enabling sharply focused beams. The only drawback is that the attenuation loss increases rapidly at high frequencies. In 5G networks, such a system is combined with small cells to overcome the attenuation loss [2, 3].

The small-length transmission lines and significant propagation losses enable spectrum reuse in cellular communications by reducing interference between neighboring cells. When longer paths are preferred, mmWave extends tiny antennas to center signals into highly focused beams with adequate gain to avoid propagation losses. Furthermore, due to the short wavelengths of mmWave signals, small multi-element, dynamic beam-forming antennas suitable for cellular communication can be built. On the contrary, the radio frequency chains in millimetre-wave substantial MIMO systems consume excessive energy and high hardware costs due to many transmit antennas. In such systems, more antennas necessitate an equal number of RF links. According to the findings, the RF components utilize up to 70% of the overall transceiver power [4, 5]. Hybrid analogue/digital transceiver systems, such as the mmWave beam-space massive (mWBSM)-MIMO design, are being studied to decrease the number of RF links.

Due to the highly directional propagation at mmWave frequencies, antenna selection and spatial modulation are the fundamental techniques of reducing the number of RF chains in MIMO systems. However, the operation is severely degraded due to the correlation of MIMO channels. Therefore, the notion of Beam-space MIMO has recently been regarded as a prominent method for decreasing the need of RF links. It utilizes the channel sparsity to convert a traditional spatial channel into a beam-space channel. The dispersed lens arrays mechanism is followed in beam-space MIMO to route signals to multiple spots on the focus surface, allowing for narrow beam width conservation and RF chain operation reduction. In addition, it allows a significant power reduction for each stream and thus reduces interference between them.

The fact that each antenna often needs its own unique RF chain and certain other related parts, is one of the most important problems with mWave massive MIMO systems. As the number of antennas grows, so does the cost of the gear and the amount of energy consumed. At mmWave frequencies, the energy consumption of an RF chain is typically substantial, around 250 mW per RF chain. However, the energy usage per RF chain is much lower at cellular frequencies, around 30 mW. As a result, for an mmWave MIMO system with 256 antennas, the RF chain energy requirement is 64 watts [6]. The energy consumption is more than the power usage in a traditional micro-cell base station. So for implementing a small number of RF chains, an mmWave massive MIMO system with a lens antenna array is found to be suitable. A lens antenna array is referred to as an antenna array matched with an electromagnetic lens, where lenses are focused on the elements of the antenna array. The spatial domain channel can be converted to a Beam-space channel by focusing the signals on different antennas. Because of the lower scattering environment at mmWave frequencies, the number of effective propagation channels in mmWave communications is lower. As a result, it occupies fewer beams, resulting in a sparse mWBS-channel. As a result, we may determine the required dominant beams based on the sparse beam-space channel for reducing significantly the size of the system and the required RF chains.

Therefore, an mmWave massive MIMO system using a lens array becomes a viable alternative for reducing energy usage. Beam selection necessitates the BS to obtain knowledge of Beam-space channel in the constrained RF configuration to get the most significant feasible performance without reducing system capacity. It can be accomplished using the mmWave channel's sparsity in the angular domain. High-resolution phase shifters realize the phase shifter network in hybrid precoding systems, allowing analogue precoders with significantly higher efficiency to be constructed to improve channel estimate accuracy. The phase shifter network is replaced with a lens antenna array in a mmWave extensive MIMO system to save hardware costs and energy [7, 8]. The channel estimation algorithms devised for hybrid precoding systems

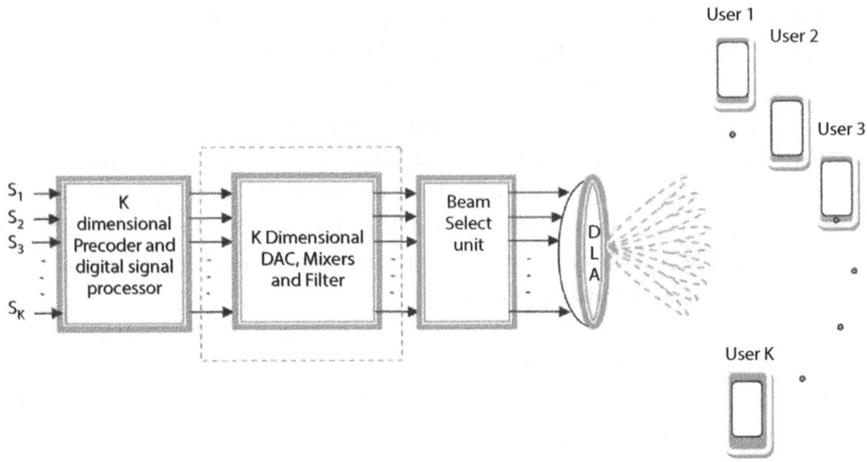

Figure 3.1 Beam-space massive MIMO system.

are sufficient to estimate the Beam-space channel for analogue precoders in the mmWave massive MIMO with a lens antenna array. BSMIMO systems have enhanced multiplexing and beam-forming capabilities for small antennas, according to recent research, and outperform comparable systems in terms of system capacity. As a result, beam selection reduces the system's RF complexity significantly. In this chapter the benefits of a mWBS-MIMO system is discussed describing the considerable improvement of beam selection using discrete lens antenna array. Moreover, the direct relationship between the number of selected beams and the radio frequency chain is achievable. The functional block diagram of the Beam-space Massive MIMO system is depicted in Figure 3.1.

3.2 Beam-Space Massive MIMO System

A devoted radio frequency chain with all other associated components is a significant problem in mmWave massive MIMO systems. The problem, in turn, induces cost increment in the hardware and, due to an extensive array of elements in the antenna, further induces a large energy consumption. From the survey [9], it is found that 250mW energy absorption in every RF chain at mmWave frequencies. But the energy absorption in individual RF chain at cellular range is comparatively lower, about 30 mW. For example, the authors [3, 4] stated that if 256 aerials are assigned in the Base Station (BS) of a mmWave massive (mWMBS)-MIMO system, then the energy absorption alone is 64 watts compared to the traditional micro-cell BS, mWMBS-MIMO that absorbs much power.

Using the lens antenna as array, the number of required RF chains in a mWMBS-MIMO system can be reduced. As said earlier, the lens antenna array can focus on the system's capacity, the signals must be focused to different antennas to modify the spatial domain into the beam-space channel. Due to a less scattering environment at mmWave frequencies, the number of actual propagation paths are less and occupies a lesser number of beams, making the mmWave beam-space channel sparse. In order to considerably decline the MIMO system size and to sparse beam-space channel, we can choose the bare minimum number of dominant beams and RF chains [10, 11]. Consequently, a lens antenna array in a large-scale MIMO system is a feasible remedy to reduce energy usage. To attain the best possible performance without degrading the system capacity, beam selection requires the BS to obtain the information on the beam-space channel in the limited RF configuration.

It can effectively be achieved through the sparsity of the mmWave channel in the angular domain. In hybrid precoding systems, high-resolution phase shifters are employed to implement the phase shifter network to advance channel estimation. In a mWMBS-MIMO system, the lens array replaces the phase shifting network that decreases cost and power consumption as well.

For analog precoders in the mmWave massive MIMO with lens antenna array, the channel estimation schemes designed for hybrid precoding systems is sufficient to estimate the beam-space channel. The analogue precoders designed are discrete Fourier transform (DFT) vectors. The Sparsity Mask Detection (SMD) estimate is preferred in mWMBS-MIMO systems incorporating lens antenna array. The main factor is finding the beams with ample power to be used.

The dimension-reduced beam-space channel depends on the beam training process of the BS and the users. So the dimension-reduced channel is estimated by standard algorithms such as the least square technique. Thus, by considering the entire beam coherence time, the estimation of the beam-space channel can be achieved with low computational complication and pilot overhead. Generally, number of BS antennas and corresponding pilot symbols to scan are proportional. However, that can be reduced by developing the sparsity nature of the beam-space channel.

In addition, the recent research shows that BS-MIMO systems have improved multiplexing and beam-forming capabilities for small antennae and outperform the comparable ones in system capacity. Thus the system RF complexity is decremented considerably by beam selection algorithms.

The advantage of the minimized system attained by the beam selection is greatly improved by the use of array of isolated lens because it relates the number of selected beams and the RF chains. Near-optimal system performance can be attained with beam selection in BS-MIMO. It does not have an influence on the beam-forming properties at the transmitter.

3.2.1 System Model

Let M BS downlink antennas with M RF chains are considered to present in an mWMBS-MIMO system. Simultaneously such a system can serve K single-antenna users. A uniform linear array configuration of M antenna elements spaced $\lambda/2$ meters apart in the access point with signal wavelength λ in meters. In transmitting beam-forming, a multiple antenna system can transmit only one symbol at a time. On the other hand, numerous transmit and receive paths in a MIMO system can provide extra degrees of freedom such that more than one symbol can simultaneously be transmitted through the MIMO channel. The system throughput can be improved by employing the spatial dimension. MIMO systems achieve the spatial dimension by multiplexing numerous streams from many transmit and receive antennas. Then the downlink vector for all the MIMO system user is represented as:

$$y = H^H Ps + w \tag{3.1}$$

where $H = [h_1, h_2, \cdots h_k]$ is the channel matrix up to h_k with k=1,2,3,... K for the channel gain of M element ULA to the kth user.

There are many scattered or reflected paths between the base station and mobile station in a wireless communication scenario. The objects which disperse the wireless signal are known as scatterers. It may be any real-world object such as trees, cars, vehicles and large buildings. As a result, the signal arriving at the mobile station has multiple paths called multipath propagation. The direct path yields the line of sight component; similarly, the scatter paths provide the non-line of sight components. Both components are included to model the channel vector for the kth user. Let P be a precoding matrix of M×K and the data vector size of K×1. To remove multi-user interference, the data elements are precoded with P. The transmit power fulfils the restraint in a manner that tr $(PP^H) \leq$ p.w is the additive white Gaussian noise (AWGN) vector, while the $w_k \sim CN(0, \sigma^2 I_k)$ follow an identical independent distribution.

3.2.2 Saleh-Valenzuela Channel Model

Small scale fading induces rapid signal amplitude changes over a small distance with arbitrary phases at the receiver. The generic model for a more significant bandwidth channel is based on the clustering phenomenon. The clustering is observed in both the temporal and spatial domains. Saleh-Valenzuela channel model is a statistical one where the cluster form of multipath components (MPCs) arrival is assumed due to multiple reflections from the objects in the surrounding area of the receiver and transmitter. Poisson distributed MPC arrival with different rates with exponentially distributed inter-arrival times are considered. An independent Rayleigh distributed random variables are considered for the amplitude of the multipath components, with the corresponding phase angles in an independent uniform random distribution over the interval (0, 2π).

According to the Saleh-Valenzuela channel model, the vector for the kth user h_k is represented as

$$h_k = \alpha_k^{(0)} b(\varphi_k^{(0)} + \sum_{l=1}^{L} \alpha_k^{(1)} b(\varphi_k^{(1)} \tag{3.2}$$

Let $\alpha_k^{(0)}$ be the path gain from ULA to the receive antenna of user k, $\varphi_k^{(0)}$ is the spatial distribution and thus $\alpha_k^{(0)} b(\varphi_k^{(0)}$ denotes the line of sight (LOS) component of the kth user. The second term $\sum_{l=1}^{L} \alpha_k^{(1)} b(\varphi_k^{(1)}$ is the NLOS component of the kth user. The total number of NLOS components and the array steering vector of order N x1 is signified by L and $b(\varphi)$, respectively.

3.3 Array Response Vector

The steering vector selection confirms an antenna array's capability to isolate the sources. The main factors behind the steering vector are the phase shift between the transducers to focus the array and the propagation distance from every transmitter to the receiver. The spectral level of the signal is deduced from the antenna array concerning a reference distance from the transmitter. In each steering vector, the choice of the weighting is based on the appropriate phase shift that is accurate to focus on a particular source. M element ULA response vector is represented as

$$b(\varphi) = 1/\sqrt{M}[e^{-2\pi\varphi n}] \tag{3.3}$$

where $n \in \tau(M) = \lceil r - (M-1)/2 \rceil$ and a symmetric index set r takes the value from 0,1,2, ... M – 1. The spatial direction Ψ is defined as $(d/\lambda)\sin\theta$, where θ and d are the directions and antenna spacing at $d = \lambda/2$.

The incident angle of arriving RF wave decides the direction of propagation, and the higher signal strength occurs during antenna rotation. The time difference of arrival (TDOA) at individual elements of the array is used for direction estimation, which can be accomplished by measuring the received phase difference in the antenna array elements. The most common method to estimate the AOA of a signal is based on beam-forming. Antenna arrays implementing beam-forming techniques electronically steer the main lobe radiation in desired direction of incoming signal. The angle of incident l^{th} component to the user is represented by $\theta_k(l)$ and $\Psi k(l) = 0.5 sin_k(l)$ is the equivalent spatial frequency. It is to note that the $\theta_k(l)$ is supposed to be in uniform distribution between the interlude $\left[-\frac{\pi}{2}, \frac{\pi}{2}\right]$.

3.3.1 mmWave Beam-Space Massive (mWBSM)-MIMO System

Figure 3.2 shows mWBSM organization in the spatial domain. The channel's randomness at mmWave frequencies dominates the LOS component more than the NLOS components with high directionality. A discrete lens antenna array is employed to modify the spatial domain into a beam-space channel.

As we said earlier, DLA allows narrow beamwidth in a reduced RF chain condition and assigns little power for every path. In addition, it provides exceptionally high antenna gain to make it suited for applications.

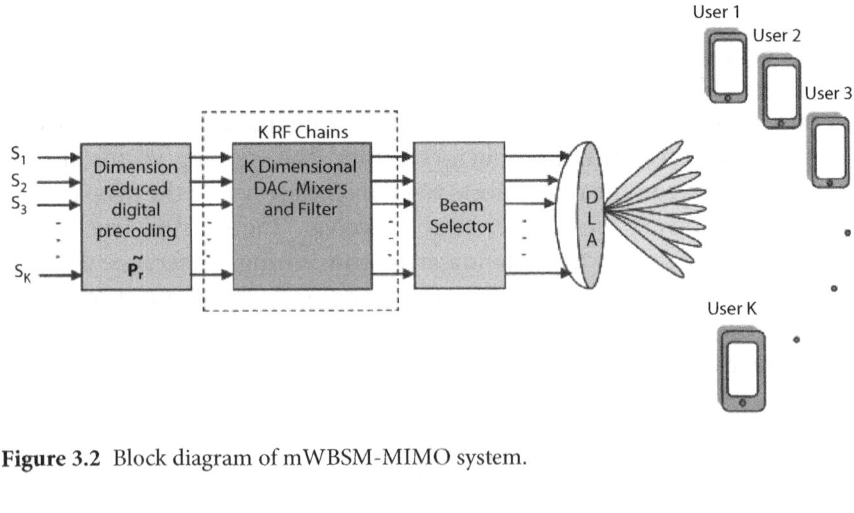

Figure 3.2 Block diagram of mWBSM-MIMO system.

The Fourier transform matrix U contains array steering vectors for space-based discrete to connect the spatial channel with the beam-space by encompassing the entire space. Let U be a unit DFT matrix that represents an ideal DLA.

$$U = [b(\mu_1), b(\mu_{12}), b(\mu_{13}), \cdots b(\mu_{1M})]^H \tag{3.4}$$

where the spatial direction is $\mu_n = 1/M\left(n - \frac{M+1}{2}\right)$ for n = 1,2,..., M and the altered received signal vector can be expressed as:

$$\tilde{y} = H^H U^H Ps + w = \tilde{H}^H P_r s + w \tag{3.5}$$

A precoder is considered in spatial multiplexing to support multi-stream transmission in MIMO wireless communication systems. Multiple signal stream transmission is needed to facilitate throughput enhancement in multiple antenna systems. The precoding can serve two purposes:

i. If the number of spatially multiplexed signals is the same as transmit antenna number, then precoding can incorporate the orthogonality behaviour of the parallel transmissions, facilitating better signal isolation at the receiver side.
ii. Precoding contributes to mapping all the spatially multiplexed signals onto the transmit antennas when fewer spatially multiplexed signals are present than the number of transmit antennas.

Based on the CSI at the transmitter, the precoder splits the input signal into spatial orthogonal beams for allocating power to them. The signal interference can be eliminated if spatial modes match the channel directions. Thus, parallel channels allow the transmission of independent data streams.

The precoder determines the radiation pattern shape for different beam-powered orthogonal beams using the CSI, which assigns more power to the vital channel directions.

3.4 Discrete Lens Antenna Array

Planar lenses are the most important part in a multiple-beam generation. A focal surface where the energy is tightly focused and the DOA of a beam are directly related. When various feeding antennas are mounted on a lens' focal surface, distinct beams are produced in the far-field area. The electronic feed excitation control can produce a variety of beam combinations. The Rotman lens is a typical illustration of a multiple-beam lens antenna while the recent design methodologies utilize lenses developed with artificial dielectric materials for multi-beam generation. Planar lenses with microstrip technology are also much suitable for this purpose. A discrete lens array comprises an array of two planar patch antennas, one of which is lighted by the feed antennas and the other by the far-field.

Like dielectric lenses, any signal can be appropriately concentrated at a particular point when the arrays are separated by a mutual ground plane and by interlocking the elements with transmission lines of appropriate lengths. The focusing zone is the focal surface, whereby feeding antennas should be positioned to increase efficiency and gain.

DLA structure has a planar and discrete configuration, making it suitable to provide finite possible paths for EM waves to travel. A microstrip line of the appropriate length is used in the array plane to implement the delay line. The DLA is more beneficial with antennas such as a parabolic dish antenna, where more significant dimensions can support the higher gain.

The parameters of a DLA are the functions of receiving antenna is located at the focal surface with distribution of power. In order to cover the focal region with appropriate angular distribution and half-wavelength spacing, the position of each receive antenna element must be selected accordingly. Thus the lens antenna array applied to a massive MIMO cellular system attains significant performance gain and reduction in cost when compared to the conventional antenna arrays.

In beam-space MIMO, the discrete lens antenna array is an essential component that serves as analog beamformer where the spatial channel is modified to the beam-space channel with insignificant degradation. It also exhibits strong structural properties in a mmWave band due to the property of sparse scattering. Furthermore, due to the consideration of each beam as a single antenna in B-MIMO, less active antennas are sufficient to obtain the desired performance.

In equation 5, $\tilde{H}$ is the beam-space channel and $Pr = U^H P \in C^{MxK}$ where P_r is the dimenstion-reduced digital precoding matrix.

$$\tilde{H} = [\,\widetilde{h_1}, \widetilde{h_2}, \widetilde{h_3}, \cdots \widetilde{h_K} = UH \in C^{MxK} \tag{3.6}$$

$$UH = [Uh_1, Uh_2, Uh_3, \cdots Uh_k] \tag{3.7}$$

where $\widetilde{h_k}$ is the beam-space channel vector of the kth user.

U represents the mapping of the signals for each UE in the spatial domain, which is obtainable from the product of the channel matrix H and the unitary DFT matrix. Each row of $\tilde{H}$ represents one of the DLA beams, where each element represents the channel gain of predefined beams. The array response vectors are taken from U's columns, matching M's orthogonal predefined directions. The objective is to select the required number of beams without any performance degradation to achieve a reduced MIMO system. Followed by the selection of beam s, the signal received at the receiver side is represented by

$$\ddot{y} = \widetilde{H_s}\widetilde{P_r}s + w \tag{3.8}$$

$\widetilde{P_r} \in C^{KxK}$ is a reduced size precoding matrix P, reflecting the reduction in the RF chain requirement. This quantity equals K to assure users' necessary spatial multiplexing gain. For maximizing sum rate in an mWB-SM-MIMO system the required number of powerful beams out of M is needed [12].

An independent and is identically distributed spatial direction $\varphi_k^{(i)}$ for i = 0,1,2,...,K ; K is with in [-1/2, 1/2] and k = 1,2, ..., is assumed. The mth element of the beam-space channel for a kth user can be signified based on equations (3.2) to (3.8). It is supposed that the ith path component in equation (3.2) is a dominant one.

$$\widetilde{h_{k,m}} = b^H(\mu_m)\sum_{i=0}^{L} \alpha_k^{(i)} b\left(\varphi_k^{(i)}\right) = \sum_{i=0}^{L} \alpha_k^{(i)}\, \gamma\left(\mu_m - \varphi_k^{(i)}\right) \tag{3.9}$$

where the function $\gamma(x) = \dfrac{\sin M\pi x}{M\, sin\pi x}$

If many users are allotted the same most substantial beam it leads to severe multi-user interference. But there exists a considerable probability for users distributed with the equal most substantial beam in B-MIMO even if the number of beams is large. If the same beam in the space is designated for dissimilar users by different RF chains, then the beam selection method is inefficient, prone to multi-user interference, and certain RF chains are being undesirably lost.

The probability of the users with the same beam can be given by

$$P = 1 - \frac{M!}{(M^K(MxK)!)} \tag{3.10}$$

For the mmWave Massive MIMO considering M = 256, K = 32 in equation (3.10) results in the 87% probability. Consequently, choosing an optimum beam for every user is desirable in practical situations.

3.5 Beam Selection Algorithm

During data transmission, a beam selection criterion identifies the beams for system. With DLA at the transmitter, the selection algorithm can be associated with the channel matrix in the beam-space domain. It does not affect the beam width or gain of the antenna. In many of the earlier works, beam selection is performed based on various constraints such as the magnitude of the path, the received SINR, system capacity, and the least bit error rate.

A maximal sum-rate objective is set using norm and uncorrelated method for choosing beam which finds K maximally uncorrelated row with the most prominent norm. An improved system capacity is obtained by applying correlation-based beam selection, and since no channel correlation of all rows is considered, it results in low complexity.

Although orthogonal channels are considered for transmission, correlated user channels induce performance degradation in real-time. So a perfect beam selection strategy is required to increase the system sum rate.

A widely used zero-forcing (ZF) precoder is implemented for beam selection in the analogue domain. To find the subset of beams used for the period of data transmission, the SINR metric must be defined for the model. Let $\tilde{P}_r$ be the precoding matrix

$$\tilde{P}_r = v\widetilde{H_s}\left(\widetilde{H_s}^H \widetilde{H_s}\right)^{-1} \tag{3.11}$$

In equation (3.11), v is a scaling factor to guarantee $E\{\|\tilde{P}_r s_2^2\| = P$, which is the transmit power of the BS.

Thus v is given as

$$v = \sqrt{\frac{P}{tr\left((\widetilde{H_s}^H \widetilde{H_s}^{-1}\right)}} \tag{3.12}$$

Based on the properties of ZF precoding, the simplified SINR of the ith user is given by

$$SINR = \frac{\gamma |v|^2}{K} \text{ where } \gamma \text{ is the signal-to-noise ratio.} \tag{3.13}$$

By assuming uniform power allocation to the users, the data rate of the kth user is given by

$$R_k = log_2\left(1 + \frac{|v|^2}{K\alpha^2}\right) \tag{3.14}$$

Thus system sum-rate is

$$R_{sum} = \sum_{k=1}^{K} R_k \tag{3.15}$$

The optimal number of beams to achieve the best the system sum-rate is given by

$$H\widetilde{_{s-opt}} = argmax_s R_{sum} \tag{3.16}$$

The beam selection algorithm selects the beam with a higher contribution to the sum rate. Therefore we can observe from the equation (3.16) that the best sum rate is equivalent to the reduced $tr\left(\left(\widetilde{H_s}^H \widetilde{H_s}^{-1}\right)\right)$.

Thus based on equations (3.12) to (3.14), equation (3.15) can be written as in equation (3.17) where $\tilde{H}(m,:)$ belongs to rows of the dimension reduced beam-space channel.

$$\tilde{H}_{s-opt} = argmin_s\ tr\left(\left(\tilde{H}(m,:)_{m\in S}^H \tilde{H}(m,:)_{m\in S}\right)^{-1}\right) \tag{3.17}$$

The beam selection algorithm obtains the optimum beam to each user having maximum sum-rate capacity by reducing the dimension of the beam-space matrix $\widetilde{H_s} \in C^{KxK}$. It can be achieved by multiplying the norm and uncorrelated rows of the beam-space channel matrix $\widetilde{H}$. Let Δ_{ij} and β_{ij} define the correlated and uncorrelated mapping between the ith and j^{th} row of $\tilde{H}$,

$$\Delta_{ij} = \frac{|h_i h_j^H|}{||h_i|| \, ||h_j||} \tag{3.18}$$

$$\beta_{ij} = \sqrt{1 - \Delta_{ij}^{\;2}} \tag{3.19}$$

where ‖·‖ and h_i are the Euclidean vector norm and the ith row of a matrix $\widetilde{H}$. At the initial stage, the largest power channel beam is selected. Then, further from the squared norm, the uncorrelated squared sum is calculated, and beams with the highest product are selected in subsequent iterations.

From that manipulations, the cost function is defined to be

$$C_{n,j} = \left\{ \begin{array}{c} ||h_j||^2 \; if \; n = 1 \\ ||h_j||^2 \sum_{k=1}^{n-1} \beta^2_{sk,j \; if \; n=2} \end{array} \right\} \tag{3.20}$$

where s_k is the kth selected beam index. For each iteration, the beam index s_k corresponds to the maximum cost function.

$$S_n = argmax_{j \in B_n} C_{n,j} \tag{3.21}$$

In equation (3.20), some sets of beam indexes are not picked and continue until n^{th} iteration. It continues until the selection of K beams from the matrix $\widetilde{H}$, where M >> K. this action makes the selection of beams with maximum power and eliminates inter-beam interference due to the greatly uncorrelated row elements of the beam-space matrix $\widetilde{H}$.

As a case study of choosing only two beams from M and beam index s_1 is selected in the first iteration, at that point $\widetilde{H}_s = \widetilde{H}(s,:)$ where $s \in s_1$. Then the determinant of $\left(\widetilde{H}_s^{\;H} \widetilde{H}_s\right)$ can be written as

$$\det\left(\left(\widetilde{H}_s^{\;H} \widetilde{H}_s\right) = \det\left(\begin{bmatrix} ||h_{s1}||^2 \; h_{s1}h_j^H \\ h_{s1}h_j^H \; ||h_j||^2 \end{bmatrix}\right)\right. \tag{3.22}$$

let us assume $\frac{1}{||h_{s1}||^2} = \mathrm{p}$ and $\frac{1}{h_j^{\;2}} = \mathrm{q}$, then $s_2 = argmin_{j \in s2} \frac{\beta^2_{s1,j}}{||h_j||^2}$ (3.23) where s_2 is the set of M-1 beam index excluding s_1. The process is continued until the selection of K beams from the beam-space matrix. This incremental process uses the correlation between the previously designated beams with the residual unselected ones at every stage. It avoids the entire correlation among the channel matrix rows. By this way a lower computational complexity is achieved with enhanced system performance at only vector multiplication. The maximal complexity is bounded to O(MK2) for selecting K beams from M.

3.6 Mean Sum Assignment-Based Beam User Association

Optimally allocating jobs to the available resources is the basic requirement in the assignment problems to obtain the needed results. Many techniques of the assignment are explained in various works. The proposed mean sum-based assignment approach is inspired by the Hungarian assignment method with the same results. However, the Hungarian method takes more space, time and changes for different problems while the proposed method is less complicated and influential in determining an optimal solution.

Let $x_{j,k}$ assign the jth beam to a kth user so as to make

$$x_{j,\,k} = 1 \; \textit{if } j^{th} \textit{ beam is assigned to } k^{th} \textit{ user} \atop 0 \; \textit{otherwise} \tag{3.24}$$

The matrix $\widetilde{H_s}$ can be configured as a cost matrix C ($i \times j$) to

$$\text{Maximize } z = \sum_{j=1}^{n} \sum_{k=1}^{n} c_{jk} x_{jk} \tag{3.25}$$

Subject to the constraints

$$\sum_{j=1}^{n} x_{jk} = 1 \text{ and} \sum_{k=1}^{n} x_{jk} = 1 \forall j, k = 1, 2, \cdots, n \tag{3.26}$$

According to equation (3.26), it means a single beam assignment to the kth user with and a jth beam to only one user. An average sum assignment method is adopted here for the beam user association compared to the Hungarian method since it is less complex to yield the optimal solution. The steps involved in the average sum assignment method are

i) Identify the highest cost value c_{ij} of the cost matrix C.
ii) Subtract the step-i) from remaining values of C.
iii) Discover if any zero is present in matrix.
iv) If the above is not satisfied, identify the minimal value in either rows or columns.
v) Deduct that minimal value from each value of the cost matrix where exist at least a zero value in each element of the matrix.
vi) In the modified cost matrix, find each zero value and calculate the mean sum value of each row and corresponding vertical column.
vii) Check each row and column having the highest mean sum and note down its index and allot specific beam to the corresponding user.
viii) An optimal beam assignment is done if the mean sum is noted at more than one place, i.e., assign the minimal value in the actual cost matrix.
ix) The corresponding row and columns are removed and moved for net iteration.

The simplex and Kuln's Hungarian methods are well known for optimal allocation problems. For example, for an (N x N) problem using the Hungarian method, the computational complexity is O(N^3), but for the above-said mean sum assignment problem, it is only O(N^3) + O(RN^2), where R is the maximum range of the cost matrix interval.

3.6.1 Performance Evaluation

An mWMBS-MIMO system with 256 antennas at the BS and 32 RF chains is arranged in a uniform linear formation. This system supports 32 users at a time with a single antenna. This system assumes one LOS and two NLOS components between the kth user and the antenna and also considered an independent path gain between the antenna and for simulation.

An independent spatial directions $\varphi_k^{(j)} \forall j$ is chosen, which covers the range of intervals of [-1/2, 1/2] in a uniform random fashion. For performance comparison with literature algorithms, the efficiency in terms of spectrum and energy is done. The energy efficiency provides the quantitative values between sum-rate performance and RF complexity. These measures help to guess the beam selection effects over the system power requirement with necessary spectrum usage.

The transmit power (p) is assumed to be 15 dBm, with energy consumption at each RF chain to be 34.4mW (P_{RF}) [4, 12]. The energy efficiency is attained as $E = \frac{R_{sum}}{P + M_{RF}P_{RF}}$ bps/Hz/W), where R_{sum} is the sum rate. Thus E is influenced not only by R_{sum} but also by the RF chains. Each RF chain provides the beam in the proposed beam-space MIMO system, reducing the required RF chain. The proposed beam selection algorithm selects the required number of dominant beams agreeing to the sparse beam-space channel to reduce the number.

Figure 3.3 and Figure 3.4 show the achieved performance measure concerning SNR (dB) for 32 and 16 users with 256 BS antennas, respectively. The simulated results are compared with Maximum Magnitude selection [13], Interference Aware Selection [14] and Greedy selection methods [15]. More users are awarded a single beam in our proposed method compared to the Maximum Magnitude selection method. So it performs well since a near-optimal QR-based decomposition approach [16] is adopted, but it is much complex. Therefore, the greedy method is very much suited for low complexity system requirements.

If we consider a complete dimensional zero-forcing algorithm, it provides an ideal upper bound performance because that method utilizes all the available RF chains. The proposed scheme achieves better performance with low complexity for various users and the minimum usage of

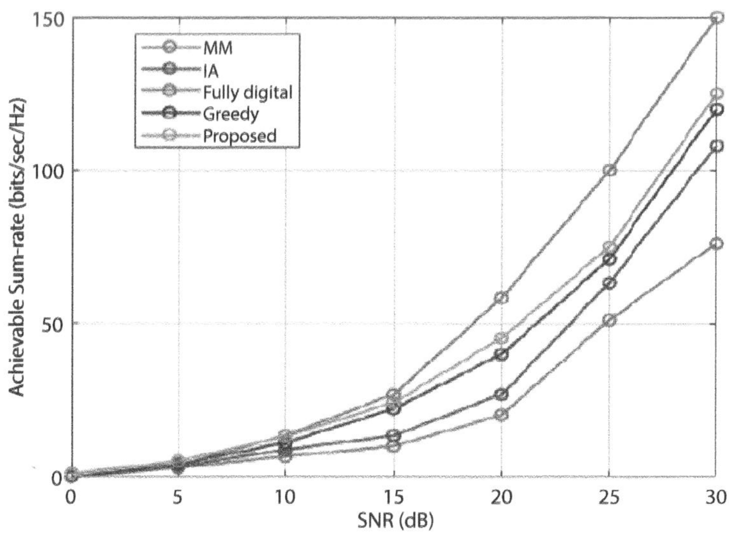

Figure 3.3 Achieved performance measure concerning SNR (dB) for 32 users with 256 BS antennas.

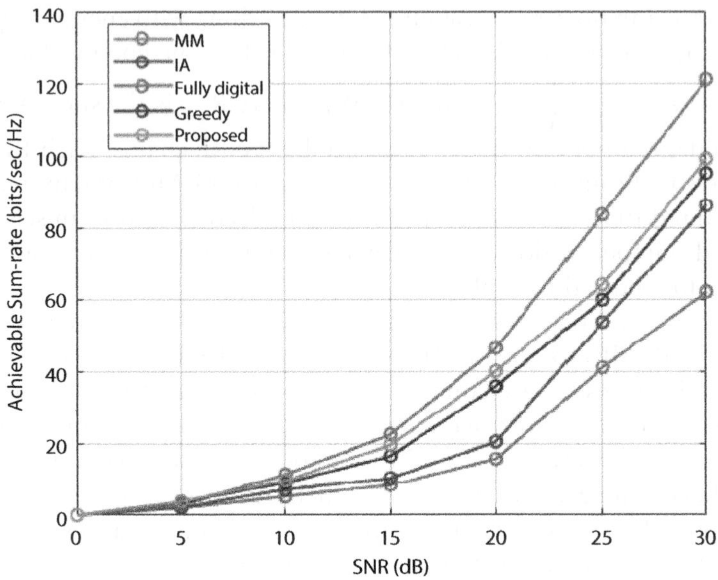

Figure 3.4 Achieved performance measure concerning SNR (dB) for 16 users with 256 BS antennas.

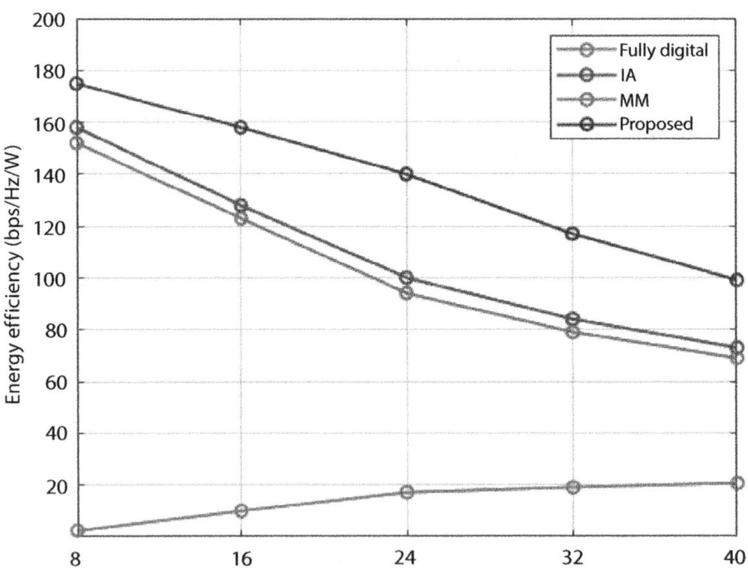

Figure 3.5 Performance in terms of energy efficiency.

RF chains. In a Greedy iterative algorithm, the signal strengths of the users are considered without considering interference between them. The proposed mean sum-rate selection method achieves a near-identical greedy approach but low complexity. Also, the proposed scheme's sum-rate performance is slightly improved under low SNR conditions.

In a full dimensional zero-forcing, many RF chains are employed to absorb more energy compared to the proposed scheme, where a minimum number of RF chains are utilized. Also proposed scheme attains the increment in spectral efficiency with low power consumption. So it improves the energy efficiency of the system also.

Each RF chain consumes lower energy of 30 mW in the cellular frequency range. So for a base station in an mmWave massive MIMO system with 256 antennas, the energy consumption of the RF chains is very high [17]. Therefore to decrease the number of required RF chains, a beam-space massive MIMO system with a beam selection strategy is considered for the proposed work, as indicated in Figure 3.5.

3.7 Conclusion

A beam-space mmWave massive MIMO system is considered where a mean sum-rate beam allocation scheme is proposed. The norm and uncorrelation approach is adopted during the beam selection scheme to maximize the sum rate in the downlink scenario. Thus the proposed beam selection with beam user association techniques for B-MIMO efficiently reduces the RF complexity of mmWave transmitters. It is inferred that improved performance is achieved in direct and interfering environments. The proposed schemes are particularly appropriate for mmWave systems because of their significance in terms of minimum usage of RF chains and power. From the simulation outcome, it is experiential that using the proposed beam selection algorithms can achieve more power efficiencies than a complete system with reduced transceiver RF complexity based on the number of mobile stations. It is also observed that comparable performance is obtained at a sum rate with an appreciable reduction in computational complication in the existing approaches.

References

1. Jiang, F., Chen, J., Swindlehurst, A.L., López-Salcedo, J.A., Massive MIMO for wireless sensing with a coherent multiple access channel. *IEEE Trans. Signal Process.*, 63, 12, 3005–3017, 2015.

2. Rappaport, T.S., Sun, S., Mayzus, R., Zhao, H., Azar, Y., Wang, K., Wong, G.N., Schulz, J.K., Gutierrez, M.S.F., Millimeter wave mobile communications for 5G cellular: It will work. *IEEE Access*, 1, 1, 335–349, 2013.
3. Rappaport, T.S., Heath, R.W., Daniels, R.C., Murdock, J.N., *Millimetre Wave Wireless Communications*, 2nd Edition, pp. 500–563, Pearson Education, United Kingdom, 2015.
4. Brady, J., Behdad, N., Sayeed, A., Beam space MIMO for millimeter wave communications: System architecture, modeling, analysis, and measurements. *IEEE Trans. Antennas Propag.*, 61, 7, 3814–3827, 2013.
5. Brady, J. and Sayeed, A., Beamspace MU-MIMO for high-density gigabit small cell access at millimeter wave frequencies. *IEEE Workshop on Signal Processing Advances in Wireless Communications*, pp. 80–84, 2014.
6. Masouros, C., Sellathurai, M., Ratnarajah, M., Large-scale MIMO transmitters in fixed physical spaces: The effect of transmit correlation and mutual coupling. *IEEE Trans. Commun.*, 61, 7, 2794–2804, 2013.
7. Liang, L., Xu, W., Dong, X., Low-complexity hybrid precoding in massive multiuser MIMO systems. *IEEE Wirel. Commun. Lett.*, 3, 6, 653–656, 2014.
8. Lu, C., Wang, W., Zhong, W., Gao, X., User scheduling and beam allocation for massive MIMO systems with two-stage precoding. *IEEE 27th Annual International Symposium on Personal, Indoor, and Mobile Radio Communications*, pp. 1–6, 2016.
9. Cui, S., Goldsmith, A.J., Bahai, A., Energy-efficiency of MIMO and cooperative MIMO techniques in sensor networks. *IEEE J. Sel. Areas Commun.*, 22, 6, 1089–1098, 2004.
10. Han, S., Xu, C., Chih-Lin, I., Rowell, C., Large-scale antenna systems with hybrid precoding analog and digital beamforming for millimeter wave 5G. *IEEE Commun. Mag.*, 53, 1, 186–194, 2015.
11. Yilmaz, A. and Kucur, O., Error performance of joint transmit and receive antenna selection in two hop amplify-and-forward relay system over Nakagami-m fading channels. *IEEE International Symposium on Personal, Indoor and Mobile Radio Communications*, pp. 2197–2203, 2010.
12. Cheng, Z., Huang, Y., Jing, Y., Jin, S., Yang, L., Sum-rate analysis for massive MIMO downlink with joint statistical beam forming and user scheduling. *IEEE Trans. Wirel. Commun.*, 16, 4, 2181–2194, 2017.
13. Sayeed, A. and Brady, J., Beam space MIMO for high-dimensional multiuser communication at millimeter-wave frequencies, in: *Proceedings of IEEE GLOBECOM*, pp. 3679–3684, 20132013.
14. Gao, X., Dai, L., Sayeed, A.M., Low RF-complexity technologies to enable millimeter-wave MIMO with large antenna array for 5G wireless communications. *IEEE Commun. Mag.*, 56, 4, 211–217, 2018.
15. Pal, R., Sarawadekar, K.P., Srinivas, K.V., Low-complexity beam selection algorithms for millimeter wave beam space MIMO systems. *IEEE Commun. Lett.*, 23, 4, 768–771, 2019.

16. Pal, R., Sarawadekar, K.P., Srinivas, K.V., A decentralized beam selection for mmWave beam space multi-user MIMO systems. *AEU-Int. J. Electron. Commun.*, 111, 4, 152884, 2019.
17. Amadori, P. and Masouros, C., Low RF-complexity millimeter-wave beam space-MIMO systems by beam selection. *IEEE Trans. Commun.*, 63, 6, 2212–2222, 2015.

Part II
OSCILLATOR & AMPLIFIER

4

Gain-Bandwidth Enhancement Techniques for mmWave Fully-Integrated Amplifiers

Shalu C.[1]*, Shakti Sindhu[2] and Amitesh Kumar[3]†

[1]National Centre for Flexible Electronics, Indian Institute of Technology, Kanpur, India
[2]Oorja Technical Services Pvt. Ltd., Indore, India
[3]Nextgen Adaptive System Laboratory (NASL), Department of Electrical Engineering, National Institute of Technology, Patna, Bihar, India

Abstract

Herein, filter fundamentals and basic design methodologies in order to attain the large gain bandwidth and additional methodology the Bode-Fano (B-F) limit have been discussed. The first section 4.1 deals with the fundamental RLC band-pass filter. In this section quality factor of filter and the noise are concisely reminded in order to set basis of resonant circuits. These circuits are generally amplifiers and oscillators for mmWave application. The next section 4.2 presents fourth order filters intended to attain gain-bandwidth improvement over the classical RLC tank. The main focus of next section is transformer based resonators. The parasitic interwinding capacitance consequence has been discussed that provides instinct on the operation of circuit. Further this conversation is stretched to attain impedance transformation in order to understand the power dividers and combiners.

Keywords: Bandpass filters, RLC tank circuits, capacitively coupled resonators, Bode-Fano limit, magentically coupled resonators, parasitic capacitance, equalization, frequency response

**Corresponding author*: shalu268@gmail.com
†*Corresponding author*: amitesh.ee@nitp.ac.in

Sangeeta Singh, Rajeev Kumar Arya, B.C. Sahana and Ajay Kumar Vyas (eds.) RF Circuits For 5G Applications: Designing with mmWave Circuitry, (55–72) © 2023 Scrivener Publishing LLC

4.1 RLC Tank

4.1.1 RC Low-Pass (LP) Filter

The basic filter is RC low-pass filter. The schematic of this is shown in Figure 4.1 and the admittance of the circuit can be written as [1–3]

$$Y = sC + 1/R = (1+sCR)/R \tag{4.1}$$

And the impedance is simply written as Z = 1/Y. It depicts a low-pass behavior with a single pole at $\omega_p = 1/(RC)$.

Here, presence of resistor is the main cause of the thermal noise and can be written as

$$I_n^2 = 4TK_B/R \tag{4.2}$$

The output present due to this noise is an output noise $\overline{V}^2n$, which is molded by the filter transfer function. The larger filter capacitance reduces the total integrated noise TK_B/C. The filter quality factor is written as

$$Q = \frac{Im(Z)}{Re(Z)} = \frac{Im(Y)}{Re(Y)} = \omega RC \tag{4.3}$$

4.1.2 RLC Band-Pass (BP) Filter

Simply introducing an inductor L, the LP RC filter is converted in to a BP filter with center frequency, $\omega_o = 1/(\sqrt{LC})$ as shown in the Figure 4.2. Final RLC BP filter with the output noise is shown in the Figure 4.2. Instinctively, at f_o, L and C resonate and the resistance of the tank decreases to R_T. Hence, the expression of output noise at ω_o is written as

$$\overline{V}_n^2(\omega) = (4TK_B)/R_T * (R_T)^2 = 4TK_BR_T \tag{4.4}$$

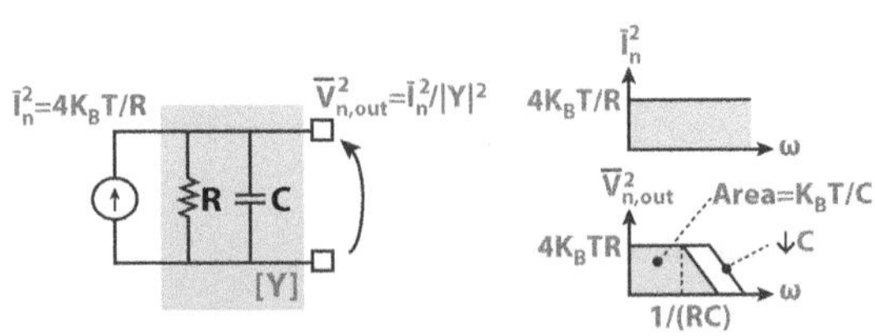

Figure 4.1 The schematic of RC low pass filter and the noise spectrum [1].

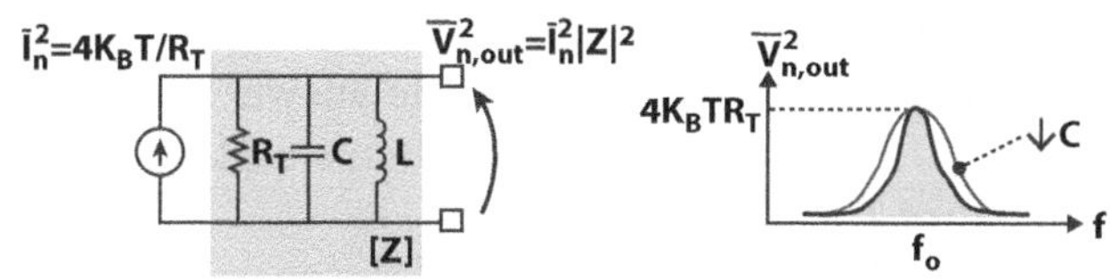

Figure 4.2 Schematic of RLC BP filter and the noise [2].

The associated noise decreases inversely with capacitor. The filter quality factor (Q) is given as

$$Q = \frac{\text{Im}(Z)}{\text{Re}(Z)} = \omega_0 R_T C \tag{4.5}$$

An oscillator can be efficiently modelled in a way as the current source parallel to RLC tank. Hence, it is worthwhile to expect to a high Q-factor tank resulting an advantage of lower noise. An amplifier can be efficiently modelled as a voltage dependent current source with a parallel RC output impedance. By inserting an inductor the amplification of the signal at mmWave is probable. In another study, load RC product bounds −3 dB is attainable bandwidth ($BW_{-3\ dB}$). The feasible approach to enhance bandwidth in the study is by introducing an equivalent resistance in shunt and reducing the tank Quality factor on expanse of additional losses, hence lesser efficiency and larger noise.

4.2 Coupled Resonators

4.2.1 Bode-Fano (B-F) Limit

Generally, mmWave CMOS amplifier is designed as ideal transconductance G_m with shunt RC I/P and O/P impedances, $R_{in}||C_{in}$ and $R_o||C_o$, respectively. In order to attain the necessary gain, noise, I/P match, O/P power characteristics, it is required to resonate the filters at parallel capacitance [1, 4]. This requires the transformation of impedance on the essential bandwidth. In this way, LNAs, PAs, on-chip gain stages and buffers have almost same design issues and their solutions up to some extent. As previously discussed, a normal way to make the capacitor resonate can be achieved by adding inductor in parallel, which led to 3-dB bandwidth limitations through the RC product of tank. There are certain doubts viz. firstly whether it's promising to resonate in a huge BW precisely. Secondly is there any theoretical limit of the problem? The last one is there any theoretical optimum solution exists or not. The answer of these problems is Bode-Fano limit [4, 5]. A passive lossless filter, shunt RC ended load with the input impedance Z_{in}.

The reflection coefficient is a measure of closeness of the input impedance with resistive termination R in a given frequency range and given as

$$\Gamma_{in}(f)=\frac{Z_{in}-R}{Z_{in}+R} \tag{4.6}$$

The Bode-Fano phenomenon emphasizes that

$$\int_{0}^{+\infty}\ln\left(\frac{1}{|\Gamma_{in}(w)|}\right)d\omega\leq\frac{\pi}{RC} \tag{4.7}$$

It is worthwhile to note that the reflection coefficient magnitude and in-band ripple are associated with each other as

$$|\Gamma_{in}|=\sqrt{1-\frac{1}{Ripple}} \tag{4.8}$$

This simple result implies that a wider pass-band bandwidth can be attained for a given RC load but at the cost of higher ripple [6, 7]. Additionally, capacitance C resonates perfectly at the certain frequencies and higher Quality factor circuits are tougher to tune in comparison to lower Quality factor circuit. In addition, RC product of transistor doesn't change with width for a specific technology at a given finger length [8]. A transistor with large width can be formulated with more fingers in shunt as well as multiple shunt transistors. This leads enhancement in shunt I/P, O/P capacitances proportional to width and a reduction in the corresponding shunt R_{in}, R_o proportional to 1/W. The Chebyshev filter is in close approximation to the ideal band pass filter.

The fourth order coupled resonators provides larger gain-B.W. in comparison to the modest RLC tank at mmWave, hence real work is focused on such realization.

4.2.2 Capacitively Coupled Resonators

The coupling of two RLC resonators is done with C_C capacitor. When $R_1 = R_2 = R$, $C_1 = C_2 = C$ and $L_{C1} = L_{C2} = L_C$, the admittance parameters of this two-port network are

$$Y_{11} = Y_{22} = \frac{1}{R} + \frac{1}{sL_C} + s(C + C_C) = \frac{1}{R}\left[1 + Q\left(\frac{s}{\omega_o} + \frac{\omega_o}{s}\right)\right],$$

$$\omega_o = \frac{1}{\sqrt{L_C(C + C_C)}} \tag{4.9}$$

$$Y_{21} = Y_{12} = -sC_C = -\frac{s\,k_c\,Q}{\omega_o\,R} \tag{4.10}$$

And Quality factor is

$$Q = \frac{R}{\omega_o L_C} = \omega_o R(C + C_C), \quad k_C = \frac{C_C}{C + C_C} \tag{4.11}$$

The transimpedance of a 2-port network is written as

$$Z_{21} = \frac{-Y_{21}}{Y_{11}Y_{22} - Y_{12}Y_{21}} = \frac{s^3 k_C Q \omega_o R}{\left[Q(1 + k_C)s^2 + s\omega_o + Q\omega_o^2\right]\left[Q(1 - k_C)s^2 + s\omega_o + Q\omega_o^2\right]} \tag{4.12}$$

Let's assume high Q-factor, complex poles of Z_{21} are evaluated

$$\omega_L = \frac{1}{\sqrt{L_C(C + 2C_C)}}, \omega_H = \frac{1}{\sqrt{L_C C}} \tag{4.13}$$

A larger C_C permits larger BP B.W. at expenditures of enhanced Q-factor of network and in-band ripple.

4.2.3 Inductively Coupled Resonators

Another method to couple RLC tanks is by the means of inductor L_c.

When $R_1 = R_2 = R$, $C_1 = C_2 = C$ and $L_{L1} = L_{L2} = L_L$, the admittance parameters of 2-port network is inscribed as

$$Y_{11} = Y_{22} = \frac{1}{R} + \frac{L_{LC} + L_L}{s\,L_L L_{LC}} + sC = \frac{1}{R}\left[1 + Q\left(\frac{s}{\omega_o} + \frac{\omega_o}{s}\right)\right] \tag{4.14}$$

$$Y_{21} = Y_{12} = -\frac{1}{s\,L_{LC}} = -\frac{\omega_o k_L\,Q}{s\,R} \tag{4.15}$$

$$Q = \frac{R(L_L + L_{LC})}{\omega_o L_L L_{LC}} = \omega_o R\,C, \omega_o = \frac{1}{\sqrt{\frac{L_L L_{LC}}{L_L + L_{LC}}C}}, k_L = \frac{L_L}{L_{LC} + L_L} \tag{4.16}$$

The transimpedance of the 2-port network is evaluated by the equation

$$Z_{21} = \frac{\omega_o^3 k_L Q\,R\,s}{\left[Q\,s^2 + s\omega_o + Q(1+k_L)\omega_o^2\right]\left[Q\,s^2 + s\omega_o + Q(1-k_L)\omega_o^2\right]} \tag{4.17}$$

Two complex poles of Z_{21} are evaluated by assuming high quality factor as

$$\omega_L = \frac{1}{\sqrt{L_L C}}, \omega_H = \frac{1}{\sqrt{\frac{L_L L_{LC}}{2L_L + L_{LC}}C}} \tag{4.18}$$

High BP B.W. can be achieved by choosing a low value of L_{LC} on expense of large in band ripple [9].

4.2.4 Magnetically Coupled Resonators

The coupling of two RLC tanks is done by magnetic coupling of the transformer,

Let us assume $R_1 = R_2 = R$, $C_1 = C_2 = C$ and $L_{M1} = L_{M2} = L_M$ and adopting the Y-parameter model of the transformer. The admittance parameters of this 2-port network is written as

$$Y_{11} = Y_{22} = \frac{1}{R} + \frac{1}{sL_M\left(1-k_M^2\right)} + sC = \frac{1}{R}\left[1+Q\left(\frac{s}{\omega_o}+\frac{\omega_o}{s}\right)\right] \quad (4.19)$$

$$Y_{21} = Y_{12} = \frac{k_M}{sL_M\left(1-k_M^2\right)} = \frac{k_M\omega_o Q}{s\,R}, \quad Q = \frac{R}{\omega_o L_M\left(1-k_M^2\right)} = \omega_o R\,C, \quad \omega_o = \frac{1}{\sqrt{L_M\left(1-k_M^2\right)C}} \quad (4.20)$$

The transimpedance of the 2-port network is written as

$$Z_{21} = \frac{-\omega_o^3 k_M Q\,R\,s}{\left[Q\,s^2 + s\omega_o + Q(1+k_M)\omega_o^2\right]\left[Q\,s^2 + s\omega_o + Q(1-k_M)\omega_o^2\right]} \quad (4.21)$$

Two complex poles of Z_{21} is written as

$$\omega_L = \frac{1}{\sqrt{L_M(1+|k_M|)C}}, \omega_H = \frac{1}{\sqrt{L_M(1-|k_M|)C}} \quad (4.22)$$

Higher magnetic coupling coefficient k permits higher BP B.W. but enhanced Q-factor of filter and in-band ripple.

4.2.5 Magnetically and Capacitive Coupled Resonator

The 2 RLC tanks are coupled by magnetically and capacitively both. Further, the investigation can be critically simplified by adopting the Y parameter model and assuming $R_1 = R_2 = R$, $C_1 = C_2 = C$ and $L_{MC1} = L_{MC2} = L_{MC}$. The admittance parameters of the filter are calculated as follow

$$Y_{11} = Y_{22} = \frac{1}{R} + \frac{1}{s\,L_{MC}\left(1-K_{MC}^2\right)} + s(C+C_{MC}) = \frac{1}{R}\left[1+Q\left(\frac{s}{\omega_o}+\frac{\omega_o}{s}\right)\right] \tag{4.23}$$

$$Y_{21} = Y_{12} = -sC_{MC} - \frac{K_{MC}}{s\,L_{MC}\left(1-K_{MC}^2\right)} + -\frac{Q}{R}\left(\frac{sk_C}{\omega_o}+\frac{k_{MC}\omega_o}{s}\right) \tag{4.24}$$

$$\omega_o = \frac{1}{\sqrt{L_{MC}\left(1-k_{MC}^2\right)(C+C_{MC})}}, k_C = \frac{C_{MC}}{C+C_{MC}} \tag{4.25}$$

$$Q = \frac{R}{\omega_o L_{MC}\left(1-k_{MC}^2\right)} = \omega_o R(C+C_{MC}) \tag{4.26}$$

The trans impedance of the two-port network can be found as

$$Z_{21}\frac{\omega_o\,Q\,R\,s\left(k_C\,s^2 + k_{MC}\,\omega_o^2\right)}{Den}$$

$$Den = \left[Q(k_C-1)s^2 + s\omega_o + Q(1-k_{MC})\omega_o^2 \cdot \left[Q(k_C+1)s^2 + s\omega_o + Q(1+k_{MC})\omega_o^2\right]\right. \tag{4.27}$$

Let us assume that the quality factor is high and kMC < 0, the complex poles of Z21 can be evaluated as

$$\omega_L = \frac{1}{\sqrt{L_{MC}(C+2C_{MC})(1-k_{MC})}}, \omega_H = \frac{1}{\sqrt{L_{MC}(1+k_{MC})C}} \tag{4.28}$$

A larger BP B.W. is attained on expenditures of in-band ripple by enhancing k_{MC} and C_{MC}.

4.2.6 Coupled Resonators Comparison

A simple design example is considered in order to compare the aforesaid fourth order filters. Characteristic values taken for I/P and O/P impedance of Gm stage. It is realized as in 28nm bulk CMOS, $R_1 = R_o = 400\Omega$, $R_2 =$

R_{in} = 1kΩ and $C = C_1 = C_o = C_2 = C_{in}$ = 14 fF. The filters are realized to attain greater than 30% small B.W. about center frequency f_o = 80 GHz. It outcomes greater than 24GHz $BW_{-3\,dB}$. $\omega_L = 2\pi$*68 GHz and $\omega_H = 2\pi$*92 GHz. The magnetic and capacitively coupled resonators based filters, are realized to match trans-impedance magnitude of the filter. For the purpose of comparison, outcomes of this study with the trans-impedance Z_{21} of classically tuned transformer (k = 0.8). Evidently, fourth order filters display a gain-B.W. improvement in comparison to simple RLC tank [10–12].

For a provided bandwidth, inductively coupled and magnetically coupled resonators shows the lowest ripple. Capacitively coupled resonators are at utmost by B-F limit. The performance of capacitive and magnetic coupled filters attains moderate characteristics. Furthermore, on-chip capacitors at mmWave have problem of less quality factor. This reduces the every technology node.

It is worthwhile to note that transformer permits a considerable area reduction and an easy DC feed and AC coupling. In practical on-chip implementation in comparison to inductively coupled resonators. 2 filters demonstrate the precisely similar $|Z_{21}|$ as expected. This can be understood as the equivalent model of transformer is a π-network which contains three single-ended inductors/four differential inductors. The magnetic field is controlled in an improved way in transformer that makes coupling convenient to control and model. Additionally, effect of mannequins in layout is lesser crucial in the transformer. Hereafter, the transformer-based filters will be considered due to these reasons.

4.3 Resonators Based on the Transformers

4.3.1 On the Parasitic Interwinding Capacitance

The corresponding lumped element model of the same is also drawn here. The parasitic to the silicon substrate C_{ox}, C_{Si}, r_{Si}, the parasitic intra-winding capacitance C_{m1}, C_{m2} and the inter-winding capacitance C_C. Even though the model is precise over a range of bandwidth, but because of the complexity, this is generally used to calculate precise value of every component by measurement and simulation.

In this model parasitic of the substrate, intra-winding capacitances modelled as corresponding shunt RC model [13]. That makes to engage them in filter terminations. In order to make simple investigation and to acquire more details of the influence of parasitic interwinding capacitance C_C, R_{S1} and R_{S2} are ignored. The equivalent diagram of the resulting two-port filter.

The voltage across C_C is

$$V_{Cc} = V_2 - V_2 = I_1(Z_{11} - Z_{21}) = I_1\left(|Z_{11}|e^{j\angle Z_{11}} - |Z_{21}|e^{j\angle Z_{21}}\right) \quad (4.29)$$

The voltage drop at C_C will be highest if $\angle Z_{11} - \angle Z_{21} = \pm 180°$. The drop will be lowest for $\angle Z_{11} - \angle Z_{21} = 0°$. The understanding is basic to comprehend the C_C effect on filter response.

It is worthwhile to note that in a broadband design when a transformer ($k < 0$) is used, lesser L_1, L_2 and k are used to counter effect C_C, which results the additional decrease in parasitic interwinding capacitance [15]. If non-inverting transformer is employed, high value of k is required to counter effect of C_C, which further cause to increase the parasitic interwinding capacitance.

Now, let's assume $L_1 = L_2 = L$ the self resonant frequency of the transformer is

$$f_{SRF} = \frac{1}{2\pi\sqrt{2L(1-k)C_C}} \quad (4.30)$$

Comparison of the self-resonant frequencies of an inverting ($f_{SRF,\ k<0}$) and non-inverting transformer ($f_{SRF,\ k>0}$) with same L, C_C and $|k|$.

$$f_{SRF,k<0} = \sqrt{\frac{(1-|k|)}{(1+|k|)}}\, f_{SRF,k>0} \quad (4.31)$$

It is clear that $f_{SRF,k<0}$ is always less than $f_{SRF,k>0}$. Hence, accepting f_{SRF} as FOM to upper frequency limit of transformer. This may provide incorrect conclusions. So $k > 0$ is essentially better in the design of high operating frequency circuits.

4.3.2 Effect of Unbalanced Capacitive Terminations

Equations for ω_L and ω_H are derived by supposing $C_1 = C_2 = C$. Now applying the same procedure outlined in frequency of complex poles where $C_2 = nC_1$ can be derived as

$$\omega_L = \frac{1}{\sqrt{L_{C1}(C_1 + C_C(1+1/n))}} = \frac{1}{\sqrt{L_{L1}C_1}} = \frac{1}{\sqrt{L_{L2}C_2}} = \frac{1}{\sqrt{L_{M1}(1+|k_M|)C_1}}$$
$$= \frac{1}{\sqrt{L_{M2}(1+|k_M|)C_2}}$$

$$\omega_H = \frac{1}{\sqrt{L_{L1}C_1}} = \frac{1}{\sqrt{L_{L2}C_2}} = \frac{1}{\sqrt{\frac{L_{L1}L_{LC}}{L_{L1}(1+1/n)+L_{LC}}C_1}} = \frac{1}{\sqrt{L_{M1}(1-|k_M|)C_1}}$$
$$= \frac{1}{\sqrt{L_{M2}(1-|k_M|)C_2}}$$

Generally the value taken for inter stage matching network of PA. Here driver was rationalized by factor of 2. Additionally, for a low noise amplifier the extent of the amplifiers in chain is not enhanced to reduce the power [9, 12–14]. $\omega_L = 2\pi*68$ GHz and $\omega_H = 2\pi*92$ GHz are forced and the quality factor of the load is kept constant (e.g. $R_2 = 1k\Omega/2$) for in order to compare. The frequency response of magnetically coupled resonator and single tuned transformer don't change [16]. That's false for all other 4th order filters. Capacitively, inductively coupled resonators demonstrate stable behavior iff $C_1 = C_2$. In order to resolve this problem Norton transformation is applied in a four-step design method which begins at inductively coupled resonators and develops a transformer-based filter at last, has been defined. These two resonators depict the frequency response in the middle. Clearly, the design example depicts the importance of proposed design method.

4.3.3 Frequency Response Equalization

When the case of lossy inductor is taken, the filter frequency response has amplitude inequity at two resonant peak.

In order to attain flat frequency response deprived of introducing components. Another method is to eliminate capacitive load. The filter is reconstructed by disturbing values of L M1 and L M2.

Let's assume the design parameters

$$\omega_{L,H}^2 = \frac{1+\xi \pm \sqrt{(1+\xi)^2 - 4\xi\left(1-k_M^2\right)}}{2L_{M2}C_2\left(1-k_M^2\right)} \tag{4.32}$$

ξ is leveraged to gain pre-emphasis. For particular ξ, magnetic coupling coefficient defines ω_H/ω_L. Furthermore, the quality factor of inductors is relatively large at mmWave frequencies. Hence, pre-emphasis need is limited i.e. ξ approximately equal to one is enough to normalize frequency response. Hence, the transformer can be designed with

$$|k_M| = \frac{\omega_H^2 - \omega_L^2}{\omega_H^2 + \omega_L^2} \tag{4.33}$$

$$L_{M1} = \frac{1}{\omega_L^2 C_1(1+|k_M|)\sqrt{\xi}}, \quad L_{M2} = \frac{\sqrt{\xi}}{\omega_L^2 C_2(1+|k_M|)} \tag{4.34}$$

Equations 4.33 and 4.34 show a new visions of transformer design factors and filter response dependence.

It is noteworthy to balance filter response at this point with the help of simple designing methods. If a coupling capacitor is introduced that results a circuit. Moreover, capacitor loss is large at millimeter wave. Inserting capacitance to a network results more ripple for similar BP B.W. Hence, this method is not desirable at millimeter wave [17]. The later frequency equalization method has been associated alongside with projected mechanism, to enhance the point further. The maximum doubtful supposed value of quality factor is 10 for inductors at 80 GHz. These losses have been generally modelled with series resistor. Less quality factor consequences the low transimpedance gain. It simultaneously allows the filter design with a higher BP B.W. for same ripples.

4.3.4 On the Parasitic Magnetic Coupling in Multistage Amplifiers

Negligence of the parasitic magnetic coupling in multistage amplifiers is another aspect to retain the silicon area minimum, in practical on-chip transformers layout. Hence, a few coupling mechanism are anticipated. We assume that this effect will be worsened with cascaded numerous G_m stages to attain the necessary gain at millimeter wave.

It is assumed to cascade the similar pre designed transformer-based fourth order inter-stage matching network three times, and k_{p1} and k_{p2} are introduced. The influence of different magnitude of k_{p1} and k_{p2}, when $|k_{p1}| = |k_{p2}| = 0.02$. This is a rational supposition when transformers have been designed near to one another in order to reduce silicon area.

Ideally, for comparison $k_{p1} = k_{p2} = 0$. The consequence of parasitic magnetic coupling has been sufficient and suggests, (1) to comprise this effect in EM simulations in a design of millimeter wave multistage amplifier is inevitable, (2) To limit this behavior ground or floating shields may be introduced, (3) The signs of k_{p1} and k_{p2} are chosen to advance the B.W.-ripple constraints of an ideal isolated fourth order filter [18].

4.3.5 Extension to Impedance Transformation

To develop more understanding of planned filter operation with an unconditionally pertinent on-chip inter-stage matching constructions the model. Though, this theory is to be stretched in order to realize impedance transformation for the amplifiers interfaced at the 50Ω input/output environment. This can be accomplished considering the advantageous characteristics of the transformer employed to understand fourth order filter.

This type of filter displays same frequency response precisely when understanding a 1/n impedance transformation at the cost of a $1/\sqrt{n}$ decrease of the transimpedance gain Z_{21}. Further, this explanation is generally simpler and it does not require more involved Norton transformation [1, 9]. Additionally, the effect of transformer design constraint is directly obvious on outcome of filter.

4.3.6 On the kQ Product

In this section, the k Q product consequence over filter frequency response of with transformer's insertion losses has been examined.

The kQ product is an essential design parameter for filter frequency response. The consequence of magnetic coupling coefficient over $|Z_{21}|$ of magnetically coupled resonators based filters. Centre frequency of the filter is

$$f_o = \frac{1}{2\pi\sqrt{C_1 L_1 (1-k^2)}} = \frac{1}{2\pi\sqrt{C_2 L_2 (1-k^2)}} = 80\text{ GHz} \qquad (4.35)$$

For example, $R_1 = 400\Omega$, $R_2 = 1k\Omega$ and $C_1 = C_2 = 14$ fF, and assuming lossless transformer. Extent of Z_{21} enhances with increment in magnetic coupling coefficient till it approaches to its maximum equal to $\sqrt{R_1 R_2}/2$. At f_o

$$k = \frac{1}{Q} = \frac{1}{\sqrt{Q_1 Q_2}} \tag{4.36}$$

where $Q_1 = \omega_o R_1 C_1$ and $Q_2 = \omega_o R_1 C_1$ are the Q-factors of filter terminations. To attain wide-band frequency response and to differentiate two resonant peaks the kQ product should be larger than 1. Hence, more bandwidth is attained at the cost of a more in-band ripple with the further enhancement of kQ product.

If lossy transformer has been employed in the filter it is expected the have a large magnetic coupling factor (k) which induces higher current in secondary coil [8]. This leads to less insertion loss and valid till Q-factor of network does not alter. Let's study two windings (L_1 and L_2) transformer. Inductor losses have been modelled with series resistances $R_{L1} = \omega_o L_1/Q_{L1}$ and $R_{L1} = \omega_o L_2/Q_{L2}$. The efficiency and insertion loss of this 2-port network are

$$\eta_{trasf} = \left[-1 + \sqrt{\frac{1+(kQ)^2}{kQ}}\right]^2 \quad \& \quad I\,L_{trasf} = 20log_{10}\left[\frac{kQ}{-1+\sqrt{1+(kQ)^2}}\right] \tag{4.37}$$

The variation of IL_{trasf} with the magnetic coupling coefficient at different $Q = \sqrt{Q_{L1} Q_{L2}}$. Lower k the results in larger the insertion loss. Moreover, if the quality factor is large enough, the degradation of the transformer performance is limited [11]. Hence, at the output side of a power amplifier low-loss matching network is necessary to gain high power added efficiency. So high k value is necessary for such applications.

4.3.7 Transformer-Based Power Dividers (PDs)

Generally, PDs are key components of several receiver and PA architectures. The magnetic couple resonator can implement series or shunt PDs by considering the advantageous properties of transformers. These three ports network demonstrates similar frequency outcome of a two-port network.

Though these networks are conceptually comparable to the series PDs with several benefits. Firstly, two inductors needed at port 1 are ½ to original L_1 and four times less than the parallel equivalent. This provides lesser insertion loss in the practical realization. Secondly, the symmetry point at port 1 in a series power divider is physically accessible. It provides balanced connection to power supply. This results good common mode rejection, hence series PDs generally favored.

4.3.8 Transformer-Based Power Combiners (PCs)

In order to attain the essential high power levels under lower power supply operation generally less than 1V, CMOS power amplifier operation depends on the PC methods. Extension of two-port magnetic couple resonator is done to understand three-port PCs. Even though PCs and PDs are seemed to be associated by the similar concept [12]. It is basic alteration which needs to be explained. The PCs, undertake the 2 I/P ports. These ports are compelled by two currents of similar magnitude and phase contrasting to PD. In PDs, a port is supplying power and the two O/P ports are loaded with passive network [5]. When this postulation is ineffective (i.e. $I_1 \neq I_3$), the resulting network analysis is more associated. Further, more analysis is required to realize the operation of asymmetrically driven PCs completely. Therefore the investigations of these circumstances are outside the scope of the work presented here. Further it is worthwhile to note for good characteristics at power back-off lesser PC insertion losses should be present.

4.4 Conclusion

This chapter dealt with filter design basics, Q-factor explanation of the and consequences of Q-factor on the BP response of filter. In the next section, techniques for gain-bandwidth enhancement are discussed with a start from its conceptually best parameter which is Bode-Fano limit. This is leaded with the most accurate topologies accepted in millimeter wave design of CMOS technology.

The conceptual platform has been developed for a transformers based on magnetic couple resonators. This is responsible for their excellently flat in-band behavior at a bandwidth and constructive on-chip implementation. The general expression is derived and important design examples are explained in detail in order to gain more insight. Additionally, the effect of parasitic interwinding capacitance on filter frequency response is discussed in detail. In next section, impedance transformation and PDs and PCs techniques have been discussed. These techniques are employed deprived of alternation in filter order and addition of additional mechanisms or forfeiting the BP response.

Acknowledgments

Author Shalu C. is thankful to NCflexE, IIT Kanpur for providing encouragement and necessary resources to write this chapter. Shakti sindhu is thankful to Mr. Saurabh Pishwe, M.D. Oorja Technical Sevices Pvt. Ltd for encoragement and support. Amitesh Kumar is thankful to Dept. of Electrical Engineering, National Institute of Technology, Patna for providing resources to write this book chapter. Amitesh Kumar is thankful to the Council of Scientific & Industrial Research to provide research fellowship to carry out research during his stay at the Indian Institute of Technology, Indore.

References

1. Vigilante, M. and Reynaert, P., *5G and E-Band Communication Circuits in Deep-Scaled CMOS*, Springer International Publishing AG, part of Springer Nature, 2018.
2. Razavi, B., *Design of Analog CMOS Integrated Circuits*, McGraw-Hill Education, New York, 2000.
3. Lee, T.H., *The Design of CMOS Radio-Frequency Integrated Circuits*, Cambridge University Press, Cambridge, 2003.
4. Bode, H.W., *Network Analysis and Feedback Amplifier Design*, D. Van Nostrand Company Inc., Princeton, 1945.
5. Mazzanti, A. and Bevilacqua, A., On the phase noise performance of transformer-based CMOS differential-pair harmonic oscillators. *IEEE Trans. Circuits Syst. I Regul. Pap.*, 62, 9, 2334–2341, 2015.
6. Long, J.R., Monolithic transformers for silicon RF IC design. *IEEE J. Solid-State Circuits*, 35, 9, 1368–1382, 2000.
7. Vigilante, M. and Reynaert, P., 20.10 A 68.1-to-96.4GHz variable-gain low-noise amplifier in 28nm CMOS, in: *IEEE International Solid-State Circuits Conference (ISSCC)*, San Francisco, CA, pp. 360–362, 2016.
8. Bassi, M., Zhao, J., Bevilacqua, A., Ghilioni, A., Mazzanti, A., Svelto, F., A 40–67 GHz power amplifier with 13 dBm PSAT and 16% PAE in 28 nm CMOS LP. *IEEE J. Solid-State Circuits*, 50, 7, 1618–1628, 2015.
9. Fano, R.M., Theoretical limitations on the broadband matching of arbitrary impedances. *J. Frankl. Inst.*, 249, 1, 57–83, 1950.
10. Bevilacqua, A. and Niknejad, A.M., An ultrawideband CMO Slow-noise amplifier for 3.1-10.6-GHz wireless receivers. *IEEE J. Solid-State Circuits*, 39, 12, 2259–2268, 2004.
11. Pozar, D.M., *Microwave Engineering*, Wiley, New York, 2009.

12. Wang, H., Sideris, C., Hajimiri, A., A CMOS broadband power amplifier with a transformer based high-order output matching network. *IEEE J. Solid-State Circuits*, 45, 12, 2709–2722, 2010.
13. Ye, W., Ma, K., Yeo, K.S., 2.5 A 2-to-6GHz Class-AB power amplifier with 28.4% PAE in 65nm CMOS supporting 256QAM, in: *IEEE International Solid-State Circuits Conference-(ISSCC) Digest of Technical Papers*, San Francisco, CA, pp. 1–3, 2015.
14. Vecchi, F. *et al.*, A wideband receiver for multi-Gbit/s communications in 65 nm CMOS. in *IEEE J. Solid-State Circuits*, 46, 3, 551–561, 2011.
15. Li, C.H., Kuo, C.N., Kuo, M.C., A 1.2-V 5.2-mW 2030-GHz wideband receiver front-end in 0.18-μm CMOS. *IEEE Trans. Microw. Theory Tech.*, 60, 11, 3502–3512, 2012.
16. Razavi, B., *RF Microelectronics*, 2nd, Prentice Hall, New Jersey, 2011.
17. Bhagavatula, V., Zhang, T., Suvarna, A.R., Rudell, J.C., An ultra-wideband if millimeter-wave receiver with a 20 GHz channel bandwidth using gain-equalized transformers. *IEEE J. Solid-State Circuits*, 51, 2, 323–331, 2016.
18. Vigilante, M. and Reynaert, P., On the design of wideband transformer-based fourth order matching networks for e-band receivers in 28-nm CMOS. *IEEE J. Solid-State Circuits*, 52, 8, 2071–2082, 2017.

5

Low-Noise Amplifiers

Jyoti Priya, Sangeeta Singh* and Bambam Kumar

Department of Electronics and Communication Engineering, National Institute of Technology, Patna, Bihar, India

Abstract

Low noise amplifiers (LNA) are the basic building blocks of RF analog circuitry. Hence it is mandatory to design the LAN block in an optimized way to enhance the entire communication block. The basic architecture and characteristics of the LNAs for 5G networks are discussed in detail in this chapter. The designs of LNAs have been carefully considered for a variety of 5G applications. Its architecture method varies depending on whether the frequency band in question is narrowband or wideband. Various topologies for achieving better optimized circuit efficiency are discussed here. LNAs are usually found at the receiver's front end, where they absorb the antenna's input signal and amplify it with minimal noise. The most important consideration factor for the design of an LNA is the gain and noise. Different noise figures for single phase and multistage amplifiers are illustrated in this chapter.

***Keywords*:** Low noise amplifiers (LNA), RF performance, millimeter-wave (mmWave)

5.1 Introduction

MOSFET and BJT are two major types of three terminal devices both have their unique property and applications but MOSFETs are widely used electronics device especially in integrated circuits (ICs) which were fabricated on single chip. Compared to BJT MOSFET is quite small i.e. (it requires very small area on the silicon chip.), its manufacturing process is quite simple and it requires comparatively less power. Circuit designer used MOSFET widely because of all above property designer made it possible

**Corresponding author*: sangeeta.singh@nitp.ac.in

Sangeeta Singh, Rajeev Kumar Arya, B.C. Sahana and Ajay Kumar Vyas (eds.) RF Circuits For 5G Applications: Designing with mmWave Circuitry, (73–106) © 2023 Scrivener Publishing LLC

to pack large number of MOSFETs (greater than 200 million) on a single IC chip to implement VLSI circuit such as memory and microprocessor. Analog circuits are also implemented by MOSFET by using very small area and density is also high. Analog device as low noise amplifier that used for frequency range that is from DC to RF microwave frequency and so on up to wideband frequency (95 GHz). We put low noise amplifier right after antenna because antenna capture a signal which is very noisy and there are many blocks in receiver such as mixer add noise to it but low noise amplifier has very little noise of its own compare to other.so we can say that LNA does not add much noise but amplify everything.

Design of LNA is a critical component in RF receiver design. Low noise amplifier which provides maximum gain and low noise figure in building RF front end [1–4].

LNA design requirements:

i. Large gain to reduce noise of receiver.
ii. Input matching to minimize return loss (s11<-10db).
iii. Low noise figure (<-3db).
iv. Linearity to void the distortion.

LNA design involves trade-off between noise figure, linearity, power consumption and gain.

$$a.A+n_{2.}A+n_1A$$

$$SNR=\frac{a}{n_2+n_1}$$

Where, A = gain of LNA, n_1 = noise of LNA, n_2 = signal noise, a = signal magnitude. A must be large enough such that if we add n_3 that is noise of further system, it will be irrelevant. Figure 5.1 shows the LNA block depiction with various parameters.

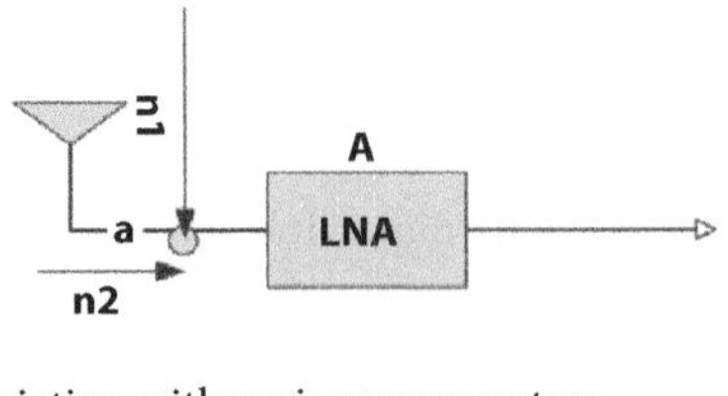

Figure 5.1 LNA block depiction with various parameters.

5.2 Basics of RFIC

5.2.1 Voltage Gain in dB

$$A_V\Big|_{dB} = 20\log\left(\frac{V_{out}}{V_{in}}\right)$$

5.2.2 Power Gain in dB

$$A_P\Big|_{dB} = 10\log\left(\frac{P_{out}}{P_{in}}\right)$$

dB_m = power in decibels referred to 1 mw. It is absolute measure.

$$p_{sig}\Big|_{dBm} = \left(\frac{p_{sig}}{1mw}\right)$$

0 dBm =1mw.
Difference between two power level is written in dB.

5.2.3 Issues in RF Design

Ideal amplifier preserves the shape of signal and we are expecting at output input should not distort. To preserve the shape of signal we assume that amplifier block is linear [5, 6]. Figure 5.2 shows the basic ideal amplifier block and Figure 5.3 illustrates the basic practical, non-ideal and non-linear amplifier block. What happened if the block is non-linear?

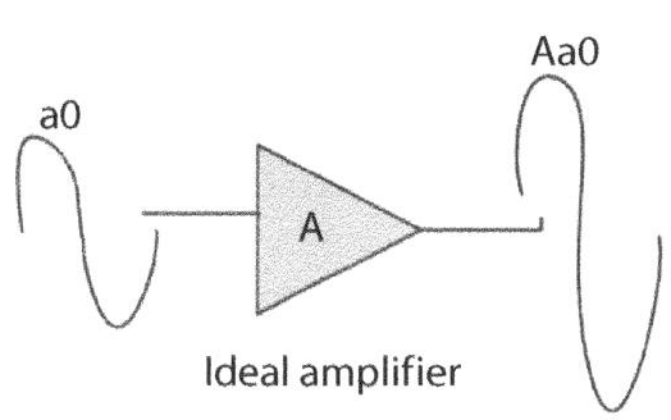

Figure 5.2 Basic ideal amplifier block.

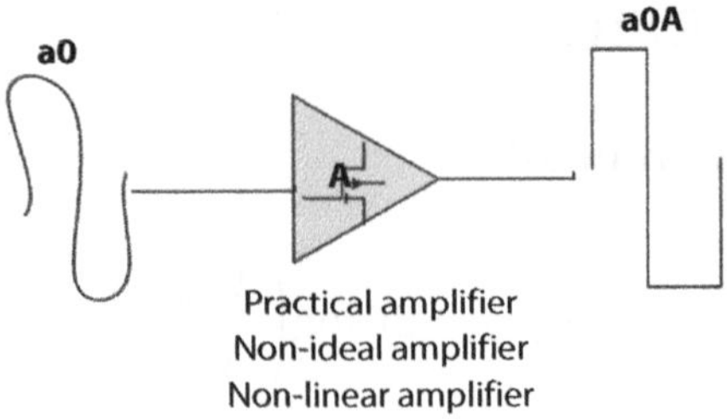

Figure 5.3 Basic practical, non-ideal and non-linear amplifier block.

There are two measure issues in RFICs are as following:

i. Non-linearity: it results in distortion.
ii. Noise: it is a small signal so it affects the minimum detectable signal in the circuit.

Non linearity in RF circuit:
For a linear system

$$y\,(t) = \alpha x\,(t)$$

α = fundamental gain
For a non-linear system:
$y(t)=\alpha_0+\alpha_1 x_1(t)+\alpha_2 x_2^2(t)+\alpha_3 x_3^3(t)+$......... this system is a memory-less non-linear.

In RF circuit to study two study the effect of non-linearity we use Two-tone-test (combination of two signals).

$$x\,(t) = A_1 \cos w_1 t + A_2 \cos w_2 t$$

$A_1 \cos w_1 t$ = desired signal
$A_2 \cos w_2 t$ = interferer signal

$$y\,(t) = \alpha_0 + \alpha_1 (A_1 \cos w_1 t + A_2 \cos w_2 t) + \alpha_2 (A_1 \cos w_1 t + A_2 \cos w_2 t)^2 + \alpha_3 (A_1 \cos w_1 t + A_2 \cos w_2 t)^3$$

Here limiting nonlinearity up to third order.

$$\alpha_0+\alpha_1 A_1 \cos w_1 t+\alpha_1 A_2 \cos w_2 t+\frac{\alpha_2 A_1^2}{2}+\frac{\alpha_2 A_2^2}{2}+\frac{\alpha_2 A_1^2}{2}\cos 2w_1 t+\frac{\alpha_2 A_2^2}{2}\cos 2w_2 t+\alpha_2 A_1 A_2(\cos(w_1+w_2)t+\cos(w_1-w_2)t)$$
$$\frac{3}{4}\alpha_3\left(A_1^3 \cos w_1 t+A_2^3 \cos w_2 t\right)+\frac{\alpha_3}{4}\left(A_1^3 \cos 3w_1 t+A_2^3 \cos 3w_2 t\right)+\frac{3A_1^2 A_2}{2}\alpha_3 \cos w_2 t+\frac{3A_1 A_2^2}{2}\alpha_3 \cos w_1 t+$$
$$\frac{3}{4}\alpha_3 A_1^2 A_2\left(\cos(2w_1+w_2)t+\cos(2w_1-w_2)t\right)+\frac{3}{4}\alpha_3 A_1 A_2^2\left(\cos(w_1+2w_2)t+\cos(w_1-2w_2)t\right)w_1\pm w_2 2w_1\pm w_2, 2w_2\pm w_1$$

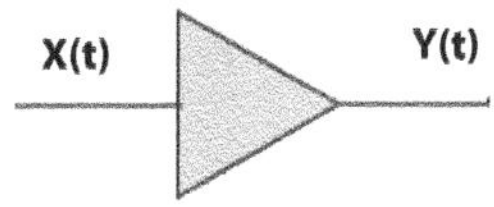

Figure 5.4 Basic amplifier block.

Effect of non-linearity:

a. Harmonics: Output of any non-linear system exhibits integer multiples of input frequency. Harmonics which are appear at the output is integer multiple of fundamental frequency. If we are dealing with narrowband signal harmonic will not create problem. For wideband signal harmonic will effect out of all harmonic component second harmonics are stronger.

$$y(t) = \alpha_0 + \alpha_1 A\cos wt + \alpha_2 A^2 \cos s^2 wt + \alpha_3 A^3 \cos^3 wt$$

$$y(t) = \alpha_0 + \alpha_1 A\cos wt + \frac{\alpha_2 A}{2} + \frac{\alpha_2 A^2}{2}\cos 2wt + \frac{3\alpha_3 A^3}{4}\cos wt + \frac{\alpha_3 A^3}{4}\cos 3wt$$

$$(\alpha_0 + \frac{\alpha_2 A^2}{2})$$

$$\left(\alpha_1 A + \frac{3\alpha_3 A^3}{4}\right)\cos wt$$

$$\frac{\alpha_2 A^2}{2}\cos 2wt$$

$$\frac{\alpha_3 A^3}{4}\cos 3wt$$

Effects of harmonics:

*i. **Harmonic distortion (HD)***

It is defined as ratio of amplitude of particular harmonic to the amplitude of fundamental component.

Figure 5.5 Basic NL system.

$$HD_2 = \frac{\frac{\alpha_2 A^2}{2}}{\left(\alpha_1 A + \frac{3}{4}\alpha_3 A^3\right)}$$

In RF circuit. HD is irrelevant indicator of the effect of non-linearity especially to narrowband system. Harmonics must be considered before it dismissed. Amplitude of nth harmonic proportional to A^n.

ii. Gain Compression

$$y(t) = \alpha_0 + \frac{\alpha_2 A^2}{2} + \left(\alpha_1 A + \frac{3}{4}\alpha_3 A^3\right)\cos wt$$

a. A is very small, than A^3 can be neglected.
b. Neglect DC also

$$y\,(t) = \alpha_1 A\cos\,wt$$

if A is significant term

$$y(t) = \left(\alpha_1 A\cos wt + \frac{3}{4}\alpha_3 A^3 \cos wt\right)$$

$\alpha_3 = +ve$ (*expensive*): Because of supply voltage limitation system has to saturate.

$\alpha_3 = -ve$ (*compressive*): It is saturating faster.

Both are non-linear because input increases but output get saturated.

1 dB compression point:

The input signal level that crosses output or gain drop by 1dB. If one dB compression point is more system is more linear.

$$20\log\left(\alpha_1 A_{1dB} + \frac{3}{4}\alpha_3 A_{1dB}^3\right) = 20\log(\alpha_1 A_{1dB}) - 1dB$$

$$\alpha_1 A_1 + \frac{3}{4}\alpha_3 A_1^3 = \frac{\alpha_1 A_1}{1.122}$$

Put $A_1 = A_{in1dB}$

$$A_{in1dB} = \sqrt{0.145\left|\frac{a_1}{a_3}\right|}$$

Desensitization:

$$x(t) = (A_1 \cos w_1 t + A_2 \cos w_2 t)$$

$A_1 \ll A_2$ (desired signal value is very-very small compared to interferer)
A_1^3 term can be neglected.

$$y(t) = \left(\alpha_1 + \frac{3}{2}\alpha_3 A_2^2\right) A_1 \cos w_1 t$$

$$\alpha_3 = -ve$$

$$y(t) \cong \left(\alpha_1 - \frac{3}{2}\alpha_3 A_2^2\right) A_1 \cos w_1 t$$

gain is decreasing function of A_2. At a large A_2: gain can turn to 0 also. Interferer blocks the desired signal by desensitizing the circuit. Hence the above phenomena are called desensitizing or blocking. If the blocking signal is greater than desired signal than the signal is desensitize.

Intermodulation:

Single signal-------------harmonic distortion
Signal + One larger interferer------------desensitization
Signal + Two large interfere---------------intermodulation

$$x(t) = A_1 \cos w_1 t + A_2 \cos w_2 t$$

$$y(t) = \alpha_0 + \alpha_1 x(t) + \alpha_2 x^2(t) + \alpha_3 x^3(t)$$

$$IM_2 = \alpha_2 A_1 A_2 \cos(w_1 + w_2)t + \alpha_2 A_1 A_2 \cos(w_1 - w_2)t$$

$$IM_3 = 2w_1 \pm w_2 = \frac{3}{4}\alpha_3 A_1^2 A_2 \cos(2w_1 + w_2)t + \frac{3}{4}\alpha_3 A_1 A_2^2 \cos(2w_2 - w_1)t$$

$$IM_3 = 2w_2 \pm w_1 = \frac{3}{4}\alpha_3 A_2^2 A_1 \cos(2w_2 + w_1)t + \frac{3}{4}\alpha_3 A_1 A_2^2 \cos(2w_2 - w_1)t$$

IM_3 falls in between the band of desired signal and output of desired signal.

Third order intercept point (IP_3):
It is used to characterize IM_2 product.

Two tone test:

$$A_1 = A_2 = A$$

Desired signal = $\alpha_1 A$

$$IM_3 \text{ component at output} = \frac{3}{4}\alpha_3 A_1^2 A_2 \cos(2w_1 + w_2)t + \frac{3}{4}\alpha_3 A_1 A_2^2 \cos(2w_1 - w_2)t$$

Since $A_1 = A_2 = A$

$$IM_3 \text{ component}: -\frac{3}{4}\alpha_3 A^3 \cos(2w_1 + w_2)t + \frac{3}{4}\alpha_3 A^3 \cos(2w_1 - w_2)t$$

A_{IP_3} performance matrix

At A_{IP_3},

Amplitude of fundamental = Amplitude of IM_3

At A_{IP_3} we can decide signal is fully corrupted.

$$|\alpha_1 A_{IP_3}| = \left|\frac{3}{4}\alpha_3 A_{IP_3}^3\right|$$

$$A_{IP_3} = \sqrt{\frac{4}{3}\frac{\alpha_1}{\alpha_3}}$$

Relation between 1dB compression point and third order intercept point:

$$\frac{A_{IP_3}}{A_{1dB}} = \frac{\sqrt{\frac{4}{3}\frac{\alpha_1}{\alpha_2}}}{\sqrt{0.145\frac{\alpha_1}{\alpha_2}}} = 3.032$$

Amplitude of A_{IP_3} in dB_m is called IIP_3.

$$\text{IIP}_3 \,|_{dBm} = P_{1dB} \,|_{dB_m} + 9.63$$

IIP_3 is almost 10 dB higher than 1dB a.

5.3 Structure of MOSFET

The MOS structure will be working as a parallel plate capacitor with metallic plate and SC substrate as two plate of the capacitor and SiO_2 as the dielectric medium. The SC region at the interface is called SC surface. The SC material away from the interface is called bulk or substrate material. The performance of MOS capacitor can be analyzed with a parallel plate capacitor as given below: Capacitor value increases after biasing due to more charge deposited on the plate [7–11].

Consider a MOS capacitor with the p-type substrate near the surface there is accumulation of positive charge so concentration of positive charge at the interface is greater than holes concentration under thermal equilibrium in p-type substrate. Hence we can say near the interface SC is more p-type. Gate is negatively biased w.r.t substrate and therefore negative charges are created on the metallic gate plate and the induced field is directed in upward direction when the field intensity penetrates the SC material a force is created on the majority carriers holes and holes are forced to moves towards the interface and they get accumulated so that an accumulation layer of positive charges or holes is created near the interface. The positive charges in accumulation layer will be now working as the positive charge for the bottom plate of the cap and the MOS capacitor is now under accumulation mode. The hole concentration near the interface will be greater than the hole concentration in the bulk and this indicates near the interface the SC is more p-type. No current will be passing through the and therefore it is under equilibrium (under equilibrium current is zero). Under strong accumulation more charge near the interface so capacitance is maximum. Let the gate is positively biased w.r.t substrate by smaller gate voltage. The gate positively biased and therefore positively charges are created on the metallic gate plate and field intensity induced and it is directed downward direction. When field intensity penetrates into the SC material. A large force created into the holes.(majority carrier of p-type) and holes are forced to move away from the interface or toward the substrate and thereby a region of immobile negative ions is created near the interface and it is called space charged region. MOS capacitor working as depletion mode. The space charge region is similar to space charge region created in the p-n junction diode. The negative charge due to immobile negative ions will be working as negative charges to the bottom plate of the capacitor. No current passing through the MOSFET. Let gate is positively biased w.r.t substrate by a larger voltage. Gate is positively biased with respect to substrate. Positive charges are created on the metallic gate plate and a larger electric field is induced and it is directed in downward direction. When larger field intensity penetrates in the SC layer. Majority carrier holes forced to move away from the interface i.e. towards the substrate and it will attract minority carrier electron towards the substrate and it will form as a layer near the interface and it is called inversion layer of electrons. Due to inversion layer of electrons near the interface the SC surface appears n-type and also the electron concentration near the interface will be larger than hole concentration in the substrate. MOS capacitor is now working under inversion mode. No current will be passing through the MOS capacitor. All possible capacitors are C_{gs}, C_{gd}, C_{gb}, $C\gg_{sb}$, C_{db} C_{gs} $\gg C_{gb}$ due to t is very small

$$C_{gb} = \frac{A\xi_{si}}{d}$$

$$C_{gd} = \frac{A\xi_{SiO_2}}{t}$$

$$C_{gs} = C_{gd} = \frac{1}{2}C_{gc}$$

As we increase drain voltage, channel will change.

$$C_{gs} = \frac{2}{3}C_{gc}$$

$$C_{gd} = \frac{1}{3}C_{gc}$$

Overlap capacitance:

$$C_{gs} + C_{ov},\ C_{gd} + C_{ov},\ C_{gd:}$$

These are small amount of capacitance but not neglected. There are some gate overlap between source and some gate overlap between drain. These overlap capacitance are independent on bias voltage they are intrinsic within the MOSFET due to fabrication. When the MOSFET is on source and drain are connected by a sheet charge or wire. Hence, no any capacitance between source and drain. Due to a wire there will be small inductance between source and drain. At junction between source and gate, and gate and drain there will be no mobile charge so it behaves as a dielectric. $C_{junction} = C_{sb}, C_{db}$ gate to body capacitance is valid only up to channel is not formed. Hence, higher the drain voltage more will be the reverse biased so capacitance will decrease further and further [12–15] C_{sb}, C_{db} function of V_{sb}, V_{db}.

Approximations:
Area of drain = A_d Area of source = A_s, perimeter of source = P_s, perimeter of drain = P_d area of junction = $A_s + P_s.h$, $A_D + P_D.h$

$$C_{jS} = \frac{(C_S + P_{Sh})C_{si}}{d_{junction}}, d_{junction} = f(V_{sb})$$

$$C_{jD} = \frac{(C_d + P_d.h)}{d_{junction}}$$

Drain is at higher potential than source in saturation so drain to body is more RB so drain to body capacitance. Gate to body is not of much concerned because they are devoided by fixed charge.

Model for MOSFET (for RF application)

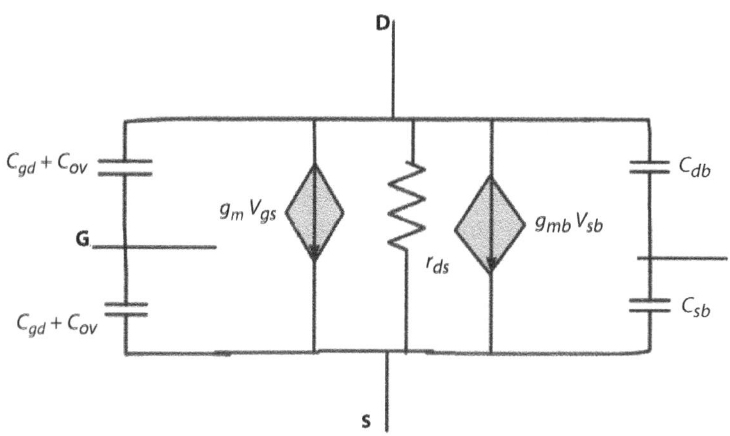

Figure 5.6 MOSFET small signal model.

5.4 Bandwidth Estimation Techniques

i. Laplace transform
ii. DC gain
iii. DC gain, poles, zeroes

$$DC\ gain = \left| \frac{\prod \left(1 + \frac{S}{S_z}\right)}{\prod \left(1 + \frac{S}{S_p}\right)} \right|^2 = \frac{1}{2}$$

Find w such that above quantity is equal to ½.

Methods of open circuit time constant:

i. It works when the system has only poles.
ii. No complex conjugate poles.

If circuit consists of resistance, capacitor, voltage controlled current source (VCCS), current controlled current source (CCCS) than we will find bandwidth, we have to pick every capacitor, open all capacitor holding outside and measure that resistance. Hence, $R_{in} \times C$ = Time constant

$$w_{-3dB} = \frac{1}{\sum \tau C}$$

Short circuit time constant:
In this method we have to know which capacitor give poles and which capacitor will give zeroes. Method of short circuit time constant: All the capacitor has to be open circuited which cause poles and all the capacitor has to be short circuit which cause zero. Open circuit time constant will give higher cut-off frequency and short circuit time constant will give lower cut-off frequency. Short circuit time constant method include only zero. This method is used in decoupling capacitor. Decoupling capacitor is large because it passes the signal at AC and blocks for DC. Hence, the decoupling capacitor cause zero and parasitic capacitor is small that will cause poles [16–21].

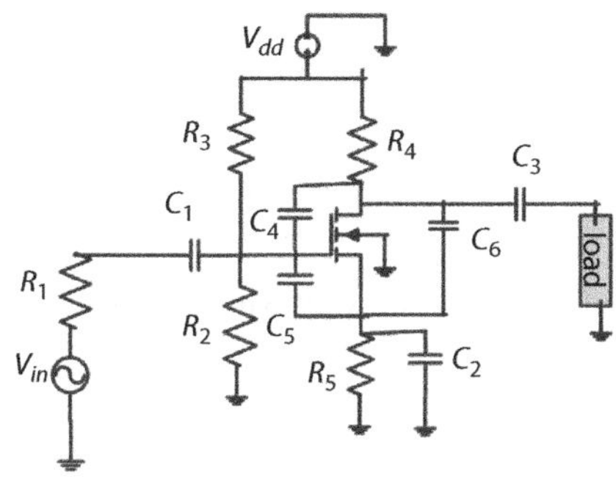

Figure 5.7 MOSFET small signal model for calculating short circuit time constant.

How can we know which capacitor contributing pole and which one zero?

If open capacitor and performance improves than that capacitor is pole contributing capacitor.

1. If open capacitor 1: If open capacitor signal will not reach to base so performance degrades due to this reason we can say that not pole contributing capacitor.
2. Open capacitor 2: If open capacitor 2 than gain will reduce so no pole contributing capacitor.
3. Open capacitor 3: If open capacitor 3 signal will not reach load so not pole contributing capacitor.
4. Open capacitor 4: If open capacitor 4 Miller effect reduces so performance improves. This capacitor is pole contributing capacitor.
5. Open capacitor 5: If open capacitor 5 less capacitance looking into base so performance improves.
6. Open capacitor 6: If open capacitor 6, capacitor 6 loading the output if we increase load, performance reduced so not pole contributing capacitor.

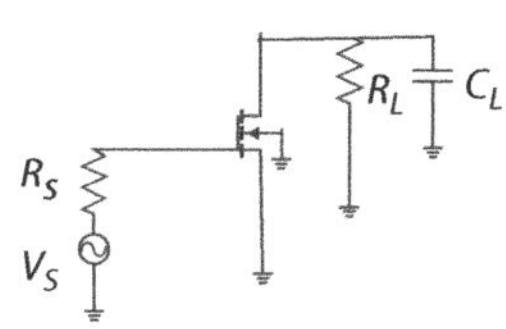

Figure 5.8 Circuit considerations for the wide band amplifier design.

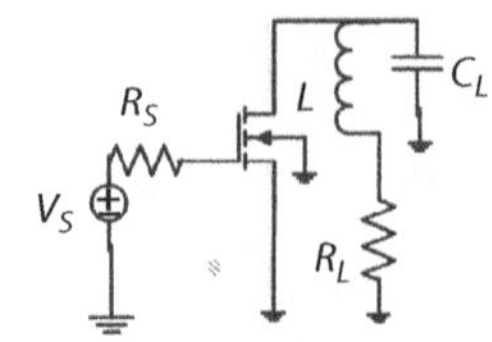

Figure 5.9 Circuit considerations for the wide band amplifier design for case-I.

Wide band amplifier design:
In earlier method of open circuit time constant bandwidth estimation we increased bandwidth by using more and more transistor. But question is how to increase bandwidth without using more and more transistor?

Case-I: For wide band amplifier using inductor in series with resistor
Impedance looking into load:

$$Z_L\,\text{earlier} = \mathrm{R}_L \left\| \frac{1}{jwC_L} = \frac{R_L}{1+jw_L R_L C_L} \text{(one pole)} \mathrm{Z}_L\,\text{now} = (\mathrm{R}_L + \text{jwL}) \right\|$$

$$\frac{1}{jwC_L} = \frac{R_L + jwl}{(1-w^2 LC_L) + jwC_L R_L} = (\textit{one}\ \text{zero and two poles})$$

$$\textit{zero}\ \text{at:} -\text{jw} = \frac{-R_L}{L}\ \textit{poles} = \frac{-R_L C_L \pm \sqrt{R_L^2 C_L^2 - 4LC_L}}{2LC_L} = \frac{-R_L \pm \sqrt{R_L^2 - \frac{4L}{C_L}}}{2L}$$

$$1.\ \textit{if}\ \mathrm{R}_L^2 \geq \frac{4L}{C_L} = \textit{pole is real.}$$

$$R_L C_L \geq \frac{4L}{R_L}$$

$$R_L C_L = \tau_C, \tau_L = \frac{L}{R_L}, \eta = \frac{\tau_C}{\tau_L}$$

$$\tau_C \geq 4\tau_L\ \eta \geq 4$$

$$2.\ \mathrm{R}_L^2 < \frac{4L}{C_L} = \textit{complex}\ \text{conjugate pole}$$

$$\mathrm{R}_L C_L < \frac{4L}{R_L}$$

$$\tau_C < 4\tau_L$$

$$\eta < 4$$

$$Z_L Now = \frac{R_L + jwL}{1 - w^2 C_L L + jwR_L C_L} = R_L \left(\frac{1 + jw\dfrac{L}{R_L}}{1 - w^2 C_L R_L \dfrac{L}{R_L} + jwR_L C_L} \right)$$

$$Z_L Now = \mathrm{R}_L \left(\frac{1 + jw\tau_L}{1 - w^2 \tau_C \tau_L + jw\tau_L} \right)$$

$$Z_L Now = R_L \left(\frac{1 + jw\dfrac{\tau_C}{\eta}}{1 - w^2 \dfrac{\tau^2}{\eta} + jw\dfrac{\tau_C}{\eta}} \right)$$

Bandwidth is defined as the frequency at which

$$|Z_L Now|^2 = \frac{1}{2}$$

$$\frac{1 + w^2 \dfrac{\tau_c^2}{\eta^2}}{\left(1 - w^2 \dfrac{\tau_C^2}{\eta}\right)^2 + w^2 \tau_C^2} = \frac{1}{2}$$

$$1 + w^2 \tau_C^2 \left(\frac{2}{\eta^2} + \frac{\tau}{\eta} - 1 \right) - \frac{w^4 \tau_C^4}{\eta^2} = 0$$

$$w^2 = \frac{\left(\dfrac{2}{\eta^2} + \dfrac{2}{\eta} - 1 \right) \pm \sqrt{\left(\dfrac{2}{\eta^2} + \dfrac{2}{\eta} - 1 \right) + \dfrac{4}{\eta^2}}}{\dfrac{2\tau_C^2}{\eta^2}}$$

$$w^2 = \frac{1}{2\tau_C^2} \left((2 + 2\eta - \eta^2) + \sqrt{(2 + 2\eta - \eta^2) + 4\eta^2} \right)$$

$$w_{-3dB}^2 = \frac{1}{\tau_C^2} \left(\left(1 + \eta - \frac{\eta^2}{2} \right) + \left(1 + \eta - \frac{\eta^2}{2} \right)^2 + \eta^2 \right)$$

η at which bandwidth is maximum:

$$\frac{df(n)}{dn}=\left((1-\eta)+\frac{1}{2}((1+\eta-\eta^2)(1-\eta^2)2(1-\eta)+2\eta)\right)$$

$$\frac{df(n)}{dn}=(1-\eta)+\frac{2+2\eta-3\eta^2+\eta^3}{\sqrt{\left(1+\eta-\frac{\eta^2}{2}\right)^2+\eta^2}}$$

$$for \text{ maximum } \frac{df(n)}{dn}=0$$

Bandwidth is maximum at $\eta=\sqrt{2}$. But peaking occurs due to complex conjugate pole are near the origin. Peaking can be avoided by pulling poles away from *jw* axis by increasing the value of η.

5.5 Noise

Noise is a random process which causes a disturbance to desired signal. Assume there is a resistor and current flowing through it so that electron moving here and there in random fashion due to temperature and at the end electron will move in direction of current flow and when electron will move around potential will develop. Hence the power spectral density of voltage developed across resistor $\left(\overline{v_n^2}\right)$ = 4KTRdf. Thus, the resistor can be modeled as:

If R = 0; Suppose ideal inductor which doesn't have any resistor. Noise voltage developed across the inductor and capacitor going to be zero. Thus inductor and capacitor are noiseless or lossless elements. If the element which is lossy i.e. consume power means convert electrical energy into heat energy which cause noise. It can be concluded that if the circuit is passive and lossless than it must be noiseless. The mean square noise voltage:

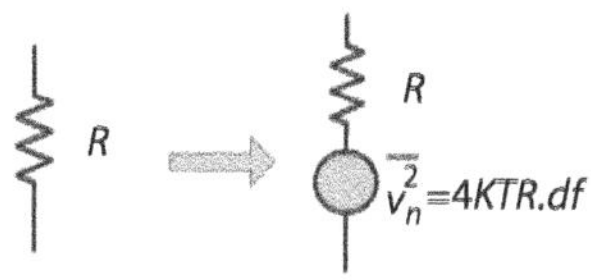

Figure 5.10 Resistor modeling for the noise modeling.

$$\left(\overline{v_{n1}+v_{n2}}\right)^2 = \overline{v_{n1}^2+v_{n2}^2+2v_{n1}v_{n2}}$$
$$\left(\overline{v_{n1}+v_{n2}}\right)^2 = \overline{v_{n1}^2}+\overline{v_{n2}^2}+\overline{2v_{n1}v_{n2}}$$
$\overline{v_{n1}v_{n2}} = 0$ ($\because$ *because* v_{n1}, v_{n2} are two uncorellated random function and its product is 0.)
$$\left(\overline{v_{n1}+v_{n2}}\right)^2 = 4\text{KTRdf} + 4\text{KTRdf} = 8\text{KTRdf}.$$

5.5.1 Noise in MOSFET

Gate noise:
Gate is made of polysilicon and it is used to make resistor as well. It means gate is resistive and it is going to contribute noise. Gate is a distributed resistor i.e. charge is distributed entirely on the gate so it cause its own lumped resistance and capacitance which cause noise. If we apply gate voltage charge will all spread on the gate since the gate material is resistive as a result we get resistance and electron moving around which can cause potential.

$$\overline{v_g^2} = 4KTR_g df$$

Channel noise:
If we apply gate to source voltage greater than threshold voltage. Channel will developed just beneath the oxide layer of the gate. Due to the formation of inversion layer electron are moving or vibrating due to temperature as these electron moved along that than it will develop potential or there will an additional noise current in shunt with the channel.

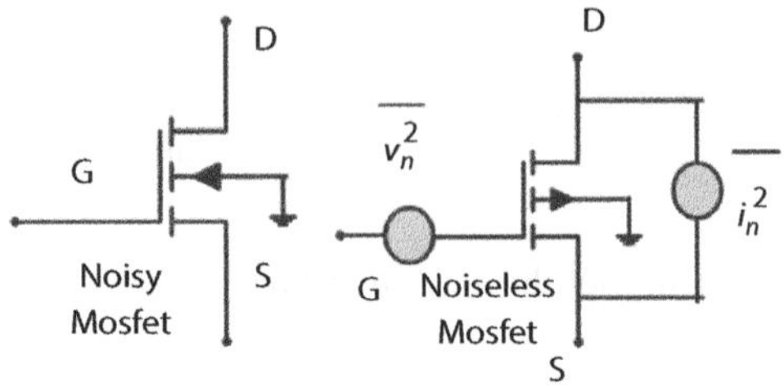

Figure 5.11 Channel noise modeling.

$$\overline{i_n^2} \propto KT$$
$$\overline{i_n^2} \propto \text{no of electron avaliabe in the channel.}$$
$$\overline{i_n^2} = KTq_I$$
$$q_I = no \text{ of electron present in the channel.}$$
$$\text{channel charge is proportional to } g_m.$$
$$q_I \propto g_m$$
$$\overline{i_n^2} = \frac{8}{3} KTg_m df$$

under condition of strong inversion or saturation. $\frac{8}{3}$ due to channel is tilted in favour of source against drain.

Body noise:
When we keep body potential equal to zero but unfortunately it will not zero Volt because of body resistance and due to this resistance there will be noise voltage. After lumping body resistance noise voltage = $4KTR_B df$. If contact resistance at gate, contact resistance at source that will also create noise.

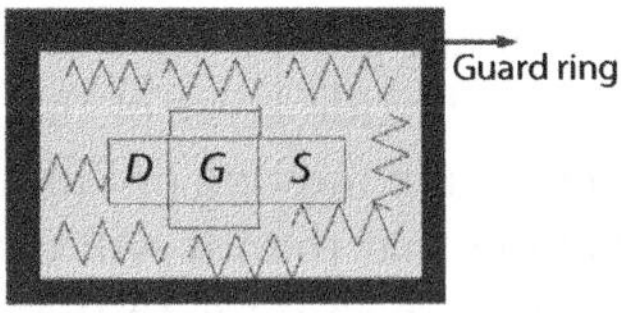

Figure 5.12 Guard ring protection approach.

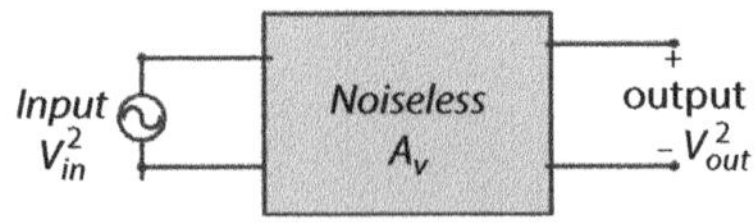

Figure 5.13 LNA modeling.

There are two definition here:

1. Output noise
2. Input referred noise

Input referred noise:
The input voltage so that we get same output noise from the noiseless system should be

$$Input \text{ reffered noise} = \frac{output \text{ noise voltage}}{gain}$$

$$i_n^2 = \frac{8}{3} KTg_m df$$

$$i_{n_r}^2 = \frac{4KTdf}{R}$$

There is no noise due to r_{ds} because r_{ds} is not a real resistor .it is due to velocity saturation in the channel because of carrier. Here by ignoring gate noise we get,

$$\overline{i_n^2}(R||\text{r}_{ds})^2 = \frac{4KTdf}{R} \frac{(R.r_{ds})^2}{(R+r_{ds})^2} = \frac{4KTdfRr_{ds}^2}{(R+r_{ds})^2}$$

$$\left(i_{n_{ch}}^2 \times (r_{ds}||R)^2\right) = \frac{8}{3} KTg_m df(R||r_{ds})$$

$$v_{nout}^2 = (r_{ds}||R)\left(\frac{1}{R} + \frac{2}{3} g_m\right) 4KTdf$$

$$gain \text{ of the circuit} = \text{g}_m \times (R||r_{ds})$$

$$v_{nin}^2 = \frac{v_{nout}^2}{g_m^2(R||r_{ds})^2}$$

$$v_{in}^2 = \frac{4KTdf}{g_m^2 R} + \frac{8}{3g_m} KTdf$$

At point 1:

$$\text{SNR} = \frac{a}{n}$$

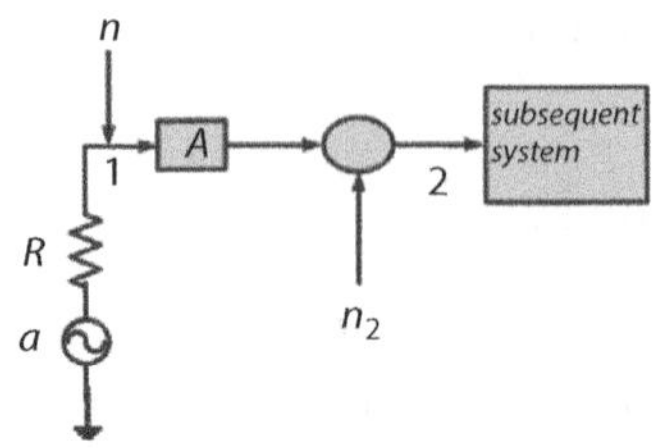

Figure 5.14 LNA modeling with resistance at input.

At point 2:

$$SNR = \frac{a.A_1}{A_1.n + n_2} \cong \frac{a}{n} (\because \text{if } A_1 \textit{ is } \text{very large.})$$

Hence, here not much effect on SNR. If gain of amplifier is high than noise of first stage is only important and all other is irrelevant. So I can ignore noise of subsequent stage and make first stage as very good amplifier. Here, we make low noise amplifier as first stage of the receiver.

5.6 Different Topologies of LNA

As we all know input impedance of MOSFET is capacitive .so to provide good matching to a 50 ohm source impedance without degrading noise performance is very difficult. We want the load must be resistive otherwise there will not matching. If there will only capacitor at load there will no maximum power transfer because capacitor will not consume power. There must be a resistor.

Gate is capacitive i.e. input impedance is capacitive so it is not possible here to do matching. We need something that dissipates the power. Solution of above problem.

Approach 1: Common source topology:
It is one of the approach to provide good matching to 50 ohm source resistance by using simply 50 ohm resistance at the gate of common source

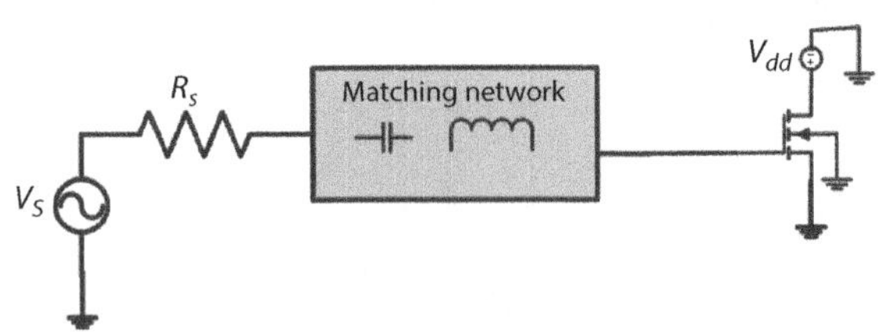

Figure 5.15 LNA modeling with matching network.

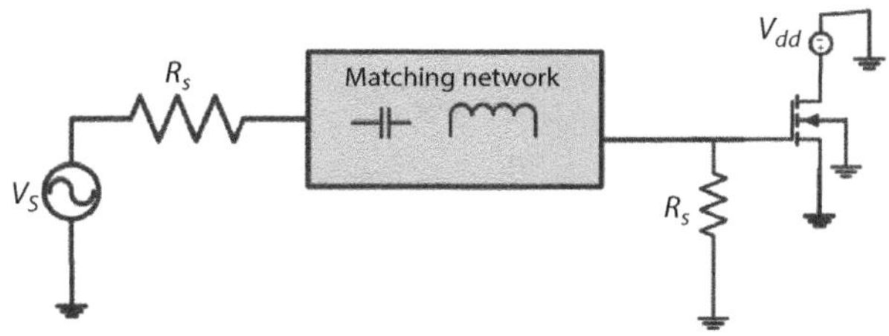

Figure 5.16 LNA modeling with matching network and resistance at output.

transistor. The resistor added at the gate of common source transistor R_s adds its own thermal noise that will attenuate the signal and also produce high noise figure. It will provide:

- Input matching
- Good gain
- Good linearity

The amount of noise achieved by antenna same amount of noise added by R_s.

For low frequency, MOSFET behaves as open circuit.

Hence, impedance = R_1, Broad band match occur at $R_1 = R_S$. Before adding amplifier we are attenuating signal by half

$$v_{in} = \frac{v_s}{2}$$

R_1 added thermal noise directly at the input.

$$F \geq 2 + \frac{4\gamma}{\alpha}\frac{1}{g_m R}, \text{ where } \mathrm{R}_1 = R_s = R.$$

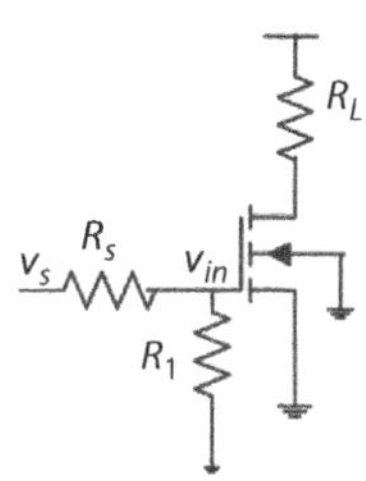

Figure 5.17 LNA modeling with resistances biases.

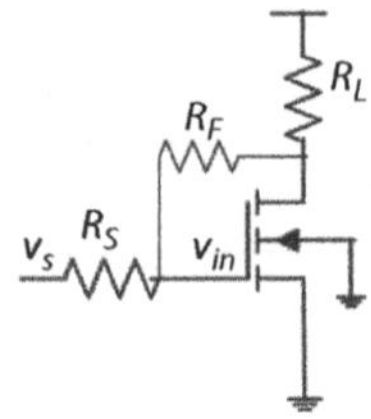

Figure 5.18 LNA designing with series shunt amplifier.

This is only applicable in low frequency and we are ignoring gate current noise. Noise figure become worse at higher frequency when we consider gate noise. Hence, the noise factor at least greater than 2 i.e. noise factor equal to 3 dB. But the noise figure of 3 dB is not good. Zeroth cut design where we do matching and nothing else did not work.

Approach 2: Shunt series amplifier:
A broadband real input impedance is provided by shunt series amplifier. It does not reduce signal strength by adding noise before amplification. Noise figure of this circuit is better than common source topology. But one drawback of this circuit is resistive feedback network continuously add noise of its own. Due to its broadband capability it is very advantageous thus the shunt series amplifier used in most of the low noise amplifier [22].

Approach 3: Common gate topology:
Hence to overcome this problem above structure is replaced by given below: For the resistive input impedance we use topology is common gate configuration. The resistance looking into source of this configuration is. $\frac{1}{g_m}$ By proper selection of device size and bias current can provide the desired input resistance [23].

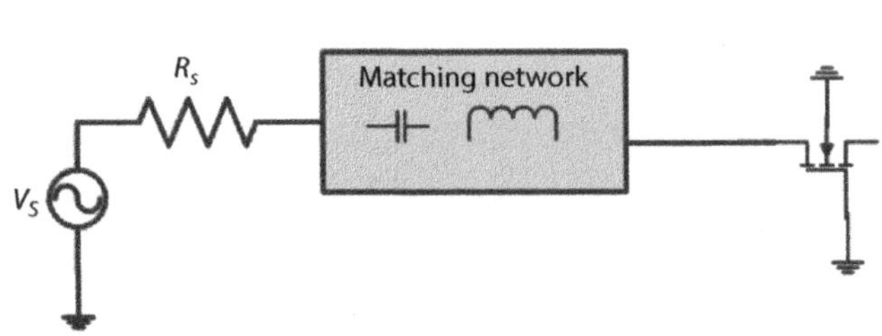

Figure 5.19 LNA designing with series shunt amplifier.

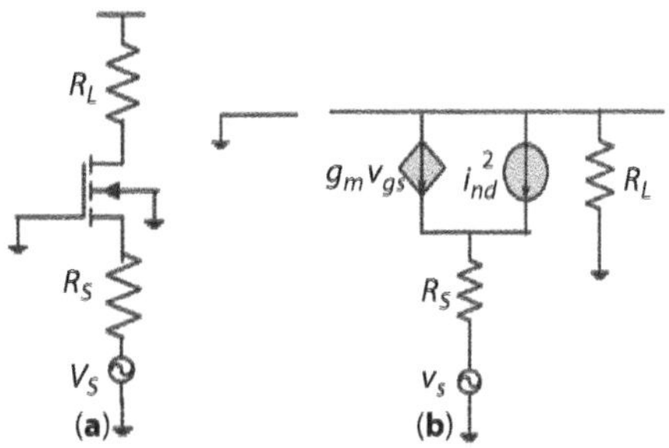

Figure 5.20 LNA designing with common gate topology.

To reduce noise factor and for proper matching the structure used here is common gate amplifier. Hence, the problem in this circuit is channel noise is in shunt with the channel and channel noise is proportional to the g_m of the device. This channel noise going to source resistance than input referred noise is proportional to g_m.

Assume $v_n R_s$ and $\overline{i_{nd}^2}$ are uncorrelated .

$$\overline{i_{nout}} = g_m v_{gs} = -g_m v_s$$

$$\overline{v_s} = \overline{v_n} R_s + \overline{i_{nout}} . R_s$$

$$\overline{v_n} : \quad \frac{-\overline{i_{nout}}}{g_m} = \overline{v_n} R_s + \overline{i_{nout}} . R_s$$

$$-\overline{i_{nout}} = g_m R_s \overline{v_n} + g_m R_s \overline{i_{nout}}$$

$$-\overline{i_{nout}} - g_m R_s \overline{i_{nout}} = g_m R_s \overline{v_n}$$

$$\overline{i_{nout}} = -\frac{g_m R_s}{(1+g_m R_s)} \overline{v_n}$$

$$\overline{i_{nd}} : \overline{v_s} = \frac{\overline{i_{nout}}}{g_m}\left(\because R_s = \frac{1}{g_m}\right)\overline{v_s} = \overline{i_{nout}} \cdot R_s \overline{i_{nout}} = -g_m v_s + \overline{i_{nd}} v_s = R_s \overline{i_{nout}}$$

$$\overline{i_{nout}} = -g_m R_s \overline{i_{nout}} + \overline{i_{nd}}$$

$$\overline{i_{nout}} + g_m R_s \overline{i_{nout}} = \overline{i_{nd}}$$

$$\overline{i_{nout}} = \frac{\overline{i_{nd}}}{1 + g_m R_s} = \frac{\overline{i_{nd}}}{2}\left(\because g_m = \frac{1}{R_s}\right)$$

$$\overline{i^2_{nout_{total}}} = \frac{\overline{i^2_{nd}}}{4} + \frac{v_n^2 R_s}{4R_s^2}$$

$$F = \frac{\overline{i^2_{nout_{total}}}}{i^2_{nout} R_s} = 1 + \frac{\frac{\overline{i^2_{nd}}}{4}}{\frac{v_n^2}{4R_s^2}} = 1 + R_s^2 . \frac{4KT\gamma g_{d0}\ f}{4KTR_s\ f}$$

$$F = 1 + \gamma g_{d0} R_s = 1 + \gamma \frac{g_{d0}}{g_m}$$

$$\alpha = \frac{g_m}{g_{d0}}$$

$$F = 1 + \frac{\gamma}{\alpha}$$

for long channel mosfet

$$\text{F} = 1.67$$

$$\text{NF} = 2.2\text{dB}$$

Here we are achieving minimum NF with common gate.

Approach 4: Inductive source degeneration:

In inductive source degeneration method the current flow in the source inductor lag by the applied gate voltage.

The advantage of this method is we can control the real part of impedance by choosing the value of inductance. This can be clarified by calculating input impedance of circuit [24, 25].

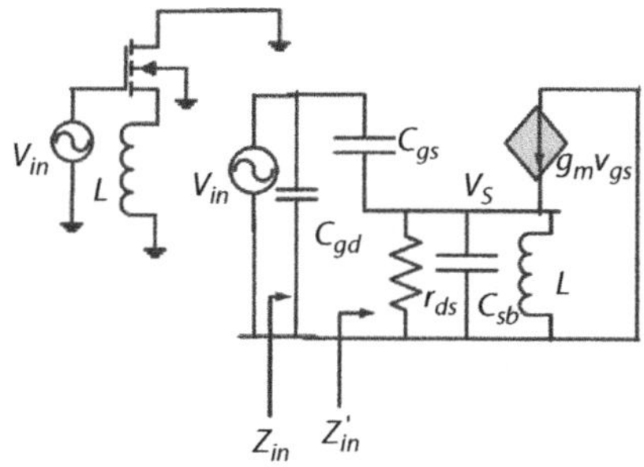

Figure 7.21 LNA designing with inductive source degeneration.

$$Z=\left(\frac{1}{jwC_{sb}}||\mathrm{r}_{ds}||\mathrm{jwL}\right)$$

$$Z_{in}=Z_{in}'\left\|\frac{1}{jwC_{gd}}\right.$$

$$KCL \text{ at node } \mathrm{v}_s(v_s-v)\,jwC_{gs}+\frac{v_s}{Z}=g_m(v-v_s)$$

$$v_s=\left(\frac{g_m+jwC_{gs}}{jwC_{gs}+\frac{1}{Z}+g_m}\right)v$$

$$I=(v-v_s)\,jwC_{gs}$$

$$I=\left(v-\frac{g_m+jwC_{gs}}{jwC_{gs}+\frac{1}{Z}+g_m}v\right)jwC_{gs}$$

$$\frac{v}{I}=Z_{in}'=\left(\frac{jwC_{gs}+\frac{1}{Z}+g_m}{jwC_{gs}.\frac{1}{Z}}\right)$$

$$Z_{in}=\left(Z+\frac{g_mZ}{jwC_{gs}}+\frac{1}{jwC_{gs}}\right)\left\|\frac{1}{jwC_{gd}}\right.$$

Case 1: Z is a resistor:

There will noise due to Z_{in}' because Z is a resistor.

Case 2: Z is a capacitor:

$$Z'_{in} = Z + Z\frac{g_m}{jwC_{gs}} + \frac{1}{jwC_{gs}}$$

$$Z'_{in} = \frac{1}{jwC} + \frac{1}{jwC}\frac{g_m}{jwC_{gs}} + \frac{1}{jwC_{gs}} = \frac{1}{jwC} - \frac{g_m}{w^2C_{gs}C} + \frac{1}{jwC_{gs}}$$

This is giving negative resistance. Negative resistance means there was some positive feedback that will cause instability hence we don't make Z as a capacitor. It makes system unstable.

Case 3: Z is an inductor:

$$Z'_{in} = Z + Z\frac{g_m}{jwC_{gs}} + \frac{1}{jwC_{gs}}$$

$$Z'_{in} = jwL + jwL\frac{g_m}{jwC_{gs}} + \frac{1}{jwC_{gs}}$$

$$Z'_{in} = jwL + L\frac{g_m}{C_{gs}} + \frac{1}{jwC_{gs}}$$

It behaves as a resistor or look like a resistor. In this circuit we are not using physical resistor but its input impedance behaves as resistor.

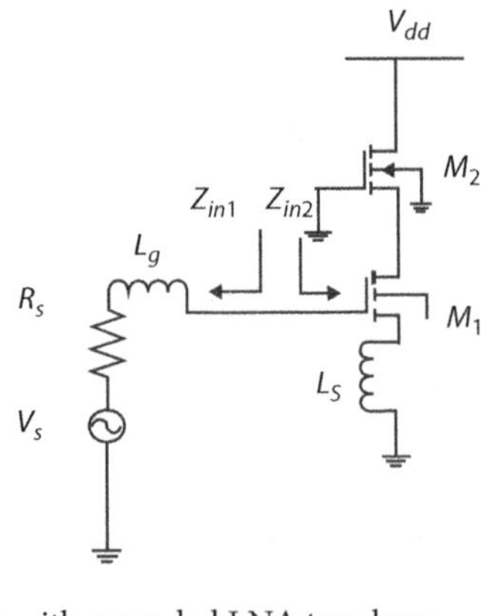

Figure 7.22 LNA designing with cascoded LNA topology.

We choose frequency such that reactive component resonate with each other and left with a resistive component only.

Approach 5: Cascoded LNA topology:

$$Z_{in1} = R_s - jwL_g$$

$$Z_{in2} = (R_s + jwL_g)^*$$

The cascoding transistor chosen here to have the same width as the main device. If it is not same than the gate drain overlap capacitance can reduce the impedance looking into the gate and drain of M_1, degrading both noise performance and input matching.

$$R_e(Z_{in}) = \frac{w_T L_S C_{gs}}{C_{gs} + 2C_{gd}}$$

Hence the degenerating inductance must be enough to compensate it. Due to miller effect produced by the common source transistor so we use a large common gate transistor as cascoded device which will reduce the gain of common source transistor so miller effect will reduced. Due to parasitic capacitance of cascoded common gate transistor at source it will lead the amplification of its own internal noise at high frequency. When we merge the source of cascoded device and drain of common source device which will reduce such kind of problem. In order to minimize the noise figure degradation, it is important to keep resistive part of input impedance large by using sufficiently large degeneration inductance to reduce miller effect. so able to reduce noise figure and also make worse effect of noise due to loss of gate inductance. Hence, to keep quality factor gate inductance as high as possible.

$$Z_S = g_{m1} r_{ds1} jwL + jwL + r_{ds1}$$

$$Z_a = g_m r_{ds} Z_S + r_{ds} + Z_S$$

C_{gd} is a small capacitance so here we are ignoring this. So impedance Z_S is primarily a huge inductor or series with an inductor. Now Z_S is shunt with a capacitor and capacitor is not large so it is not inductor. Hence again impedance looking into drain of second MOSFET is primarily an inductor.

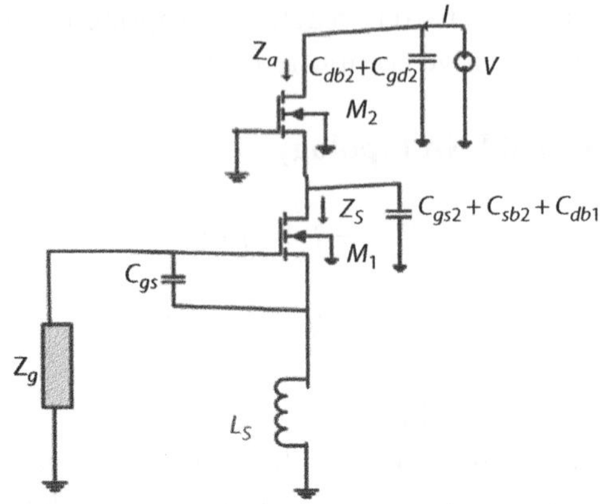

Figure 5.23 LNA designing with two MOSFET cascoded LNA topology.

It is good that impedance looking into drain of MOSFET is inductive because further we have to connect it to the gate of another MOSFET. So it is very easy to drive capacitor. Hence the inductor and capacitor resonate at certain frequency.

Approach 6: Tuned cascode LNA topology:

In above design the inductor value at drain of M_2 is not resonate with the further capacitor than we used one more inductor in shunt with that is a tuned LNA design.

Noise generating sources here is: Channel of M_1 and M_2 and gate of M_1 and M_2. First take i_{n1}^2 than null all other noise source and voltage source.

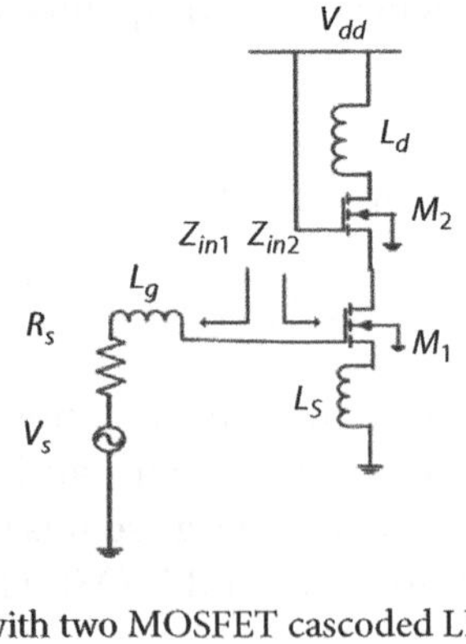

Figure 5.24 LNA designing with two MOSFET cascoded LNA topology.

Approximation 1:
Input impedance of cascode MOSFET (M_2) i.e. source of cascaded M_1 cascoded MOSFET (M_2) is 0 ohm. Voltage division happened at gate node:

$$Z_{in} = \left(jwL_s + g_m \frac{L_s}{C_{gs}} + \frac{1}{jwC_{gs}} \right)$$

$g_m \frac{L_s}{C_{gs}}$ dominates.

$$noise \text{ current} = i_n = g_m \times V_{gs}$$

$$i_2 = i_1 + g_m V_{gs}$$

$$V_g = \frac{V \times \left(\frac{g_m L_s}{C_{gs}} \| C_{gd} \right)}{\left(\frac{g_m L_s}{C_{gs}} \| C_{gd} \right) + jwL_g + R_s}$$

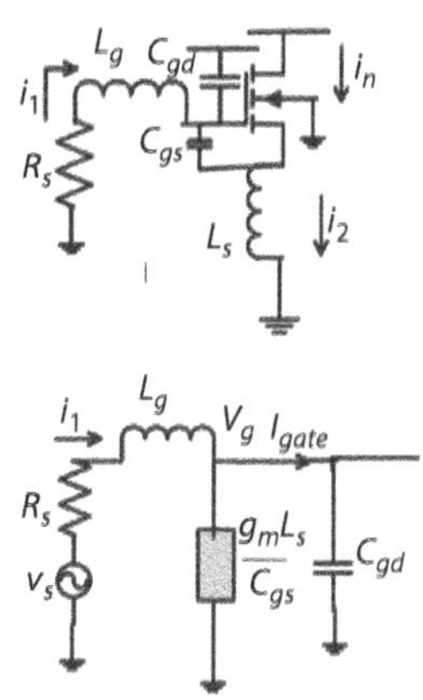

Figure 5.25 LNA designing with two MOSFET cascoded LNA topology (approximation-1).

Current going into gate:

$$i_1 = \frac{V - V_g}{\left(R_s + jwL_g\right)}$$

$$I_{gate} = \frac{i_1 \times \frac{1}{jwC_{gd}}}{\frac{1}{jwC_{gd}} + \frac{g_m L_s}{C_{gs}}}$$

$$V_{gs} = \left(\frac{V}{2}\frac{C_{gs}}{g_m L_s}\right)\frac{1}{jwC_{gs}}$$

$$i_d = \left(\left(\frac{V}{2} \times \frac{C_{gs}}{g_m L_s}\right)\frac{1}{jwC_{gs}}\right)g_m$$

$$\text{Power consume by LNA} = \frac{V^2 C_{gs}}{4 g_m L_s}$$

Now power delivered by V_s and R_s to LNA and power consume by LNA both should be equal.

$$\frac{V_s^2}{4R_s} = \frac{\left(\frac{V}{2}\right)^2 C_{gs}}{g_m L_s}$$

$$V_s = V\sqrt{\frac{R_s C_{gs}}{g_m L_s}}$$

For the certain i_n^2 we have to apply certain noise voltage to source

$$\overline{v^2} = \overline{i_n^2}.|2wL_s|^2$$

$$\overline{V_s^2} = \overline{i_n^2}.|2jwL_s|^2 \times \frac{R_s C_{gs}}{g_m L_s}$$

$$NF = \frac{total\ \text{noise}}{source\ \text{noise}} = \frac{4KTR_s df + 8KTg_m df.|2jwL_s|^2 \times \frac{R_s C_{gs}}{g_m L_s}}{4KTR_s df}$$

$$NF = 1 + 8w^2 L_s C_{gs}$$

Conclusion

The low-noise amplier givesn designing an RF front end, look for high gain and low noise. Large gain to reduce receiver noise, input matching to decrease return loss, and low noise gure are all requirements of a low noise amplifier. There are several low noise amplifier topologies, for example, a common source LNA with strong gain and input matching, but noise gure greater than 3-dB. Other topologies include common gate connected topologies so that no additional resistor is needed on the source side with the matching network employed for the purpose of matching, resulting in less noise. However, there is a problem the channel noise in this circuit is in shunt with the channel.

Acknowledgement

The authors are grateful to the National Institute of Technology, Patna's Department of Electronics and Communication Engineering for providing resources to compose this book chapter.

References

1. Gray, P.R., Hurst, P.J., Lewis, S.H., Meyer, R.G., *Analysis and Design of Analog Integrated Circuits*. John Wiley & Sons, 2009.
2. Goo, J.S., *High-Frequency Noise in CMOS Low-Noise Amplifiers*, Stanford University, 2001.
3. Steer, M., *Microwave and RF Design: A System Approach*, SciTech Pub, vol. 3, 2010.
4. Mohebi, Z., Parandin, F., Shama, F., Hazeri, A., Highly linear wide band low noise amplifiers: A literature review (2010–2018). *Microelectronics J.*, 95, 104673, 2020.
5. Li, W.T., Tsai, J.H., Yang, H.Y., Chou, W.H., Gea, S.B., Lu, H.C., Huang, T.W., Parasitic-insensitive linearization methods for 60-GHz 90-nm CMOS LNAs. *IEEE Trans. Microw. Theory Tech.*, 60, 2512–23, 2012.
6. Chen, K.H. and Liu, S., II, Inductorless wideband CMOS low-noise amplifiers using noise-canceling technique. *IEEE Trans. Circuits Syst. I Regul. Pap.*, 59, 305–14, 2012.
7. Bruccoleri, F., Klumperink, E.A.M., Nauta, B., Wide-band CMOS low-noise amplifier exploiting thermal noise canceling. *IEEE J. Solid-State Circuits*, 39, 275–82, 2004.
8. Rahman, M. and Harjani, R., A 2.4-GHz, sub-1-V, 2.8-dB NF, 475-dual-path noise and nonlinearity cancelling LNA for ultra-low-power radios. *IEEE J. Solid-State Circuits*, 53, 1423–30, 2018.
9. Yu, H., Chen, Y., Boon, C.C., Mak, P., II, Martins, R.P., A 0.096-mm2 1-20-GHz triple-path noise- canceling common-gate common-source LNA with dual complementary pMOS-nMOS configuration. *IEEE Trans. Microw. Theory Tech.*, 68, 144–59, 2020.
10. Razavi, B. and Behzad, R., RF microelectronics, 2, 255−333, New York, Prentice Hall, 2012.
11. Li, C., Li, M., He, K., Tarng, J., A low-power self-forward-body-bias CMOS LNA for 3–6.5-GHz UWB receivers. *IEEE Microw. Wirel. Compon. Lett.*, 20, 2, 100–2, 2010.
12. Chang, C.P., Chen, J.H., Wang, Y.H., A fully integrated 5 GHz low-voltage LNA using forward body bias technology. *IEEE Microw. Wirel. Compon. Lett.*, 19, 176–8, 2009.
13. Huang, B.J., Lin, K.Y., Wang, H., Millimeter-wave low power and miniature CMOS multicascode low-noise amplifiers with noise reduction topology. *IEEE Trans. Microw. Theory Tech.*, 57, 3049–59, 2009.
14. Sun, K., Tsai, Z., Lin, K., Wang, H., *A 10.8-GHz CMOS Low-Noise Amplifier Using Parallel-Resonant Inductor*, pp. 1795–8, 2007.
15. Kim, Y.M., Han, H., Kim, T.W., A 0.6-V +4 dBm IIP3 LC folded cascode CMOS LNA with gm linearization. *IEEE Trans. Circuits Syst. II Express Br.*, 60, 122–6, 2013.

16. Balamurugan, K. and Kartha, K.S., Design of LNA using body biasing techniques for enhanced reliabilty and robustness. *Proc. 3rd Int. Conf. Commun. Electron. Syst. ICCES 2018*, pp. 803–10, 2018.
17. Ponton, D., Palestri, P., Esseni, D., Selmi, L., Tiebout, M., Parvais, B., Šiprak, D., Knoblinger, G., Design of ultra-wideband low-noise amplifiers in 45-nm CMOS technology: Comparison between planar bulk and SOI FinFET devices. *IEEE Trans. Circuits Syst. I Regul. Pap.*, 56, 920–32, 2009.
18. Pan, D., Duan, Z., Chakraborty, S., Sun, L., Gui, P., A 60-90-GHz CMOS double-neutralized LNA technology with 6.3-dB NF and-10dBm P-1dB. *IEEE Microw. Wirel. Compon. Lett.*, 29, 489–91, 2019.
19. Hossain, M.Z., Hossain, M.A., Islam, M.S., Rahman, M.M. Chowdhury, M.H.., Electrical characteristics of trigate finfet. *Global Journal of Researches in Engineering, Electrical and Electronics Engineering (GJEEE)*, 11, 7, 2011.
20. Callender, S., Shin, W., Lee, H.J., Pellerano, S., Hull, C., FinFET for mmWave-technology and circuit design challenges. *2018 IEEE BiCMOS Compd. Semicond. Integr. Circuits Technol. Symp. BCICTS 2018*, pp. 168–73, 2018.
21. Shin, W., Callender, S., Pellerano, S., Hull, C., A compact 75 GHz LNA with 20 dB Gain and 4 dB Noise Figure in 22nm FinFET CMOS technology. *Dig. Pap.-IEEE Radio Freq. Integr. Circuits Symp.*, June 2018, pp. 284–7, 2018.
22. Borremans, J., Parvais, B., Dehan, M., Thijs, S., Wambacq, P., Mercha, A., Kuijk, M., Carchon, G., Decoutere, S., Perspective of RF design in future planar and FinFET CMOS. *Dig. Pap.-IEEE Radio Freq. Integr. Circuits Symp.*, pp. 75–8, 2008.
23. Lu, Y.T. and Jin, J.D., 100-mV 44-μW 2.4-GHz LNA in 16 nm FinFET technology. *IEEE MTT-S. Int. Microw. Symp. Dig*, August 1–3, 2016, 2016.
24. Yeh, E.H.V., Lo, A.-H., Chen, W.-S., Yeh, T.-J., Chen, M., *A 16 nm FinFET 0.4 V Inductor-Less Cellular Receiver Front-End with 10 mW Ultra-Low Power and 0.31 mm2 Ultra-Small Area for 5G System in Sub-6 GHz Band*, 2017.
25. Kim, Y., Jang, P., Lim, J., Ko, W., Heo, S., Lee, J., Cho, T.B., A Ka-band phase shifting low noise amplifier with gain error compensation for 5G RF beam forming array using 14nm FinFET CMOS. *Proc. IEEE Int. Symp. Circuits Syst.*, May 1–4, 2018, 2018.

6
Mixer Design

Brajendra Singh Sengar[1]* and Amitesh Kumar[2]†

[1]Department of Electronics and Communication Engineering, National Institute of Technology, Srinagar, India
[2]Nextgen Adaptive Systems Group, Department of Electrical Engineering, National Institute of Technology, Patna, India

Abstract
Several aspects of mixer designs have been discussed in this chapter. Several mixer specifications were considered. In addition, several examples were considered for the illustration. With detailed analysis of works from several research groups, diode-mixers as well as several features of transistor mixers, including their various forms, were discussed in detail. Mixers are an essential component of 5G communication architecture. Their incorporation into 5G communication architecture and subsequent implementation alongside other components will be critical for the future of 5G.

Keywords: 5G, mixer, mixer design, transceiver, SNR, diode mixer, millimeter wave communication, transistor mixer

6.1 Introduction

In recent years, Millimeter wave (mmW) communication in the 28 GHz radio frequency band has grown in importance for 5G (fifth generation) wireless communication and radar applications [1]. In the current technologies, the problem of high data rate, wider bandwidth, and low latency is prevailing. These issues are not addressed by fourth generation (4G) communication via LTE (Long Term Evolution). Therefore, more advanced communication techniques such as 5G is needed which can resolved this issue of concern [2]. In recent times, for 5G communication new radio (NR) frequency bands of 3GPP (third-generation partnership project)

**Corresponding author*: brajendra.singh@nitsri.ac.in
†*Corresponding author*: amitesh.ee@nitp.ac.in

Sangeeta Singh, Rajeev Kumar Arya, B.C. Sahana and Ajay Kumar Vyas (eds.) RF Circuits For 5G Applications: Designing with mmWave Circuitry, (107–122) © 2023 Scrivener Publishing LLC

is distributed. The frequency bands are divided into frequency ranges 1 (FR1) and 2 (FR2), which contain frequencies ranging from 0 to 6 GHz and above 24 GHz, respectively [3].

As we know, in the medical, industrial, virtual and augmented reality, and 8K video streaming applications, the demand for high data rate is increasing fast [4]. Due to increase in the use of different internet related services in tablets, internet of things, and smartphones, it is projected that wireless data traffic would increase 10,000 times in the next few years. Recently, in the automotive industry the 28 GHz frequency band radio has established much consideration owing to its probable use in autonomous vehicles [4]. Short-range radars are widely used in the automotive industry for modern intelligent transportation systems. Short-range radars are extremely safe and intelligent for self-driving cars, providing a solid foundation for safe travel [5]. For various applications, the 5G communication standard will provide ultra-high mobile broadband, low latency, high reliability, and very low energy consumption. The IoT, e-health, and high definition audio/video streaming are such applications which can explore 5G communication at larger extend [4]. The one of the attractive feature of 5G is low latency which means very less delay in communication. It becomes important in applications where quick response times is requires like medical operation theaters and autonomous vehicles.

Semiconductor devices operated at high-frequency devices at 28 GHz band having very less area are very important are for communication exploration and the research is also increased in this area. Due to low dimension of these devices, it leads to low power consumption and the device integration is increased which lead to low cost of the whole component. Devices, which are operating at 60 GHz frequency band have been, investigated thoroughly. The allocation of this band for high speed data rate, high quality video streaming, and high capacity is the reason for extensive research [6, 7].

The research effort has now shifted to the 28 GHz communication band. A modular design for 5G NR (New Radio) and FR2 (Frequency Range 2) has been proposed. These modular designs are made up of two distinct antennas. One antenna is dedicated to a microwave receiver operating at a frequency of 26.5 - 29.5 GHz, and the second antenna is dedicated to an RF conventional direct conversion receiver (DCR) operating at a frequency of 0 - 6 GHz. A switch was used to select the RF frequency for the receiver. The microwave receiver down-converts the RF frequency to an appropriate intermediate frequency (IF), which is then down-converted to a baseband frequency by the direct conversion receiver. A super heterodyne double down conversion receiver is the name given to this structure.

The goal of this book chapter is to propose new design ideas for an integrated mixer for a 5G radio transceiver front-end operating at 28 GHz in 22 nm FDSOI technology, and then to simulate these ideas. Furthermore, a comparison of the various performances of mixer topologies in terms of optimized metrics is provided. Subsequently. As we all know, the down conversion mixer is a critical component in the design of a receiver front-end [7]. The down conversion mixer limits the system's overall linearity and noise figure. The front-end receiver and down conversion mixer have been extensively researched. In 28 nm, 45 nm, and 65 nm nodes, complementary metal-oxide semiconductor (CMOS) technology was used to develop the front-end receiver and the down conversion mixer. However, no research has been conducted on fully depleted silicon-on-insulator (FDSOI) technology with 22 nm nodes at 28 GHz frequency band for down conversion mixers. Subsequently the important parameter for measuring the performance of down conversion mixer is high linearity and less sensitivity to noise. Generally, the noise of circuit's increases and linearity degrades as the technology shrinks. Therefore, the prime focus of this book chapter is to compare the linearity performance of mixer topologies.

6.2 Properties

a) Gain/Loss

The conversion gain is an important performance parameter metric for the mixer (or loss).

Conversion gain is defined as the ratio of the IF output to the value of the RF input, or more precisely, the ratio of the rms (root mean square) voltage at the IF output to the rms voltage at the RF input. It is generally preferred that a mixer's conversion gain be greater than one. It will reduce the noise contribution in the subsequent stages of the receiver front end. The active mixer's conversion gain can be expressed as a power quantity. However, the performance of a passive mixer can be better expressed in terms of voltage or current conversion gain. In most modern RF designs, voltage conversion gain is chosen. The impedances are not always equal and mostly imaginary, power quantities make it difficult to use for RF design, which is why voltage conversion gain is preferred.

Because active mixers provide amplification as well as frequency translation, their conversion gain is typically greater than one. Since the technology node is being phased out, the supply voltage is decreasing. To maintain the same linearity, this downsizing limits the conversion gain of active mixers. Passive mixers have a conversion gain that is less than -3 dB and cannot be greater than unity. As a result, passive mixers outperform active mixers in terms of linearity. A down conversion mixer's voltage conversion gain was measured by applying a sinusoidal signal at the RF port and measuring the signal amplitude at the IF port.

b) Noise Figure

Noise figure has been used to describe the quality of a signal at the receiver front end (NF). NF is defined as the ratio of the signal-to-noise ratio (SNR) at the input of the (RF) port to the SNR at the output (IF) port. The noise figure can be defined for both single and double sidebands (SSB and DSB). It is necessary to understand the process of down conversion mixing before defining noise figures for SSB and DSB. The mixer takes two frequencies as input and outputs an intermediate frequency. The desired RF signal is one of these two signals, and the unintended image signal is the other. Furthermore, the intermediate frequency is the amplitude of the difference in frequencies between the RF and LO. The input signal, the noise present in the input signal band and the noise in the image band of the mixer has been decodes to the intermediate frequency (w_{if}). The mixer gives $w_{if} = w_{rf} - w_{lo}$ and $w_{if} = w_{image} - w_{lo}$ frequencies. Thus, the IF contains the desired RF signal and the undesired image signal.

The noise figure is defined as the SSB noise figure if the desired signal occurs only on one side of the local oscillator (LO) frequency. However, if the desired signal is present on both sides of the LO frequency, the quantity is known as the DSB noise figure. A mixer's SSB noise figure is typically 3 dB higher than its DSB noise figure. As a result, SSB and DSB have the same IF noise. However, in SSB, the signal power exists only on a single sideband, implying that the output SNR is half that of the input SNR (considering same power of both noise power). Noise figures in amplifiers are generally lower than noise figures in mixers because noise from frequencies other than the desired RF frequency can mix down to the intermediate frequency. Furthermore, because the mixer is a noisy component, a low noise amplifier (LNA) is used to amplify signal levels far above the noise of the mixer and subsequent stages.

Linearity

Linearity is a critical performance metric. The size of the signal for the system is defined by linearity. It establishes the upper limit for the input signal and indicates how large a signal can be used. The output of a perfect linear system varies linearly with the input of the system. A linear system that is static in nature and cannot be determined by past input or output values (a system without memory) can be defined as:

$$V_{out} = \alpha\, w_{IN}$$

Where V_{out} is the output of the system, v_{IN} is the input of the system, and α is the constant which is function of time for time-variant system.

If the IF output is proportional to the magnitude of the RF input signal, the mixer is a linear system. However, in practice, the transfer function is not so simple. It is very complicated due to a variety of factors such as passive and active constituents in the system or a power supply that limits the signal swing. A power series can be used to express the input-output relationship of a nonlinear system mathematically.

Nonlinearity allows infinite terms to be perfectly defined. However, for most real-world applications, the first three terms are sufficient. In general, the output equations include DC terms, harmonics of fundamentals, and intermodulation products of those harmonics. The most important components are first-order mixing terms. In mixer design, all other higher order terms are unnecessary. Second- and third-order harmonic distortion (HD2 and HD3), second- and third-order intermodulation terms (IM2 and IM3), and intermodulation of harmonic components are examples of undesirable spectral components. Third-order products, on the other hand, are the intermodulation distortion products in the frequency domain. This intermodulation distortion product consists of the fundamental signals as well as the second harmonics of the other signals in the system. Many of the other harmonics are out-of-band and do not cause any issues. The third-order intermodulation products $2\omega1 \pm\omega2$ and $2\omega2 \pm\omega1$ are, however, close to the fundamental frequency. This band may fall within the intermediate frequency passband, causing distortion to the output signal.

Third-order intercept point (IP3) is another linearity figure of merit. It is used to define a mixer's intermodulation performance. The linearity of a mixer can be described using two-tone third-order intermodulation products. It was necessary to apply two signals with equal amplitudes at the mixer's input, normally one channel apart. The graph required to determine the IP3 point as a function of input power level between the desired or fundamental output and third-order intermodulation (IM3) output. In general, fundamental output rises by 20 decibels per decade, while IM3 output rises by 60 decibels per decade. The slopes of the fundamental output and the IM3 output are one and three, respectively. The intersection point (an imaginary point) can be found by extrapolating the two curves. The intersection point determines the input referred third-order intercept point (IIP3) and the output referred third-order intercept point (IIP3) (OIP3).

A second-order intercept point (IP2) is also an important parameter for measuring performance. It also defines the design of the direct down conversion receiver.

It is caused by second-order distortion products and is well-defined in the same way that the third-order intercept point is. When the curves are inferred, the second-order intermodulation terms rise at a rate of 40 dB/decade and intersect the fundamental output.

d) Interport isolation

Port-to-port isolation between the mixer's three ports is critical because signal leak from one port to another can corrupt the desired signal. Signal leakage occurs due to capacitances present in the MOSFET device, such as gate to drain and gate to source capacitance. However, the degradation of mixer performance due to port-to-port feed-through is dependent on the architecture used.

LO-to-RF Feedthrough: The LO signal is a large signal that can leak to the RF port, reach the antenna, and radiate from the antenna in a direct down-conversion receiver, which is very undesirable. Because LO leakage can add to the RF input and then be translated to IF, it can also cause DC offset at the mixer's output. With symmetric mixer design (including devices) and LO waveforms, the problems of LO radiation and DC offset at the output can be eliminated.

Offset is also known as LO self-mixing because the leaked LO signal is added to the RF input signal and mixed with the LO waveform. When the source resistance is reduced to zero, the output DC offset disappears as well. The LO-to-RF leakage is unimportant in the case of a heterodyne receiver because the LO is outside the RF band and is attenuated by the LNA, the band-select filter, and the antenna.

RF-to-IF Feedthrough: In both direct and reflected light, the RF to IF isolation is typically high. The architecture of a conversion receiver and a heterodyne receiver. RF-to-IF leakage, on the other hand, can contaminate the baseband signal due to the beat component caused by even order the presence of distortion in the RF path.

LO-to-IF Feedthrough: In direct conversion receivers, the LO to IF feed-through can cause issues for the stage after the mixer; however, a low-pass filter is usually used to pass the desired frequency while blocking all high frequencies. If the if ωIF and ωLO of the second stage mixer are very close to each other in a heterodyne architecture, the LO feed-through can desensitize the following mixer (causing DC offset in the baseband).

e) Port VSWR

VSWR (Voltage Standing Wave Ratio) is a critical parameter in mixer design. The mixer is regarded as an essential circuit that is used in front of the communication system. At transmitter end, it functions to translate the signals which are information, from low to high frequency, whereas at receiver end it does the reverse [8]. Return loss and standing wave, which are thought to be the same index, can be converted into each other. VSWR has been considered in the wo The voltage standing wave ratio (VSWR) is a critical parameter in mixer design. A mixer is a critical circuit that is used in front of a communication system. Its function at the transmitter end is to translate informational signals from low to high frequency, while it does the opposite at the receiver end [8]. Return loss and standing wave, both of which are thought to be one index, can be converted into one another. VSWR has been taken into account in the work of Ding *et al.* [9]. VSWR is typically less than 2 or 2.5 in engineering practice. In engineering practice, the VSWR should be less than 2 or 2.5.

f) DC offset

DC-offset is a significant contributor to leakage, as well as causing noise figure and affecting receiver linearity. Direct-conversion (DC) radio receivers can significantly reduce silicon area consumption. These DC receiver systems have image frequency issues, but they are hampered greatly by DC-offsets. Consider RF mixers, where linearity is critical because it affects the overall system's dynamic range [10]. The VGA topology, of the work [11], consists of the following components: i) a high-pass filter, ii) a four-stage VGA core, iii) a DC offset cancellation unit, and iv) an output buffer. The high-pass filter was used to remove the input DC offset produced by the receiver [11]. Furthermore, LO leakage is one of the causes of DC offset in mixers. Symmetric mixer and LO waveform designs may significantly reduce this issue. Offset is also known as LO self-mixing because a leaked LO signal can add to the RF input and mix with the LO waveform. If the source resistance approaches zero, the output DC offset can be eliminated [3].

g) DC polarity

Not only would DC offset be a critical factor in mixer design, but so would DC polarity [3]. For a double-balanced mixer, as illustrated in this design [3] Bashir *et al.* have illustrated a design where Local Oscillator (LO) have been used to feedthrough introduced by LO+ and LO-. Their diametrically opposed DC polarity aids in cancelling out feedthrough at the output end. As a result, the conversion gain of a double-balanced mixer is half that of a single-balanced mixer. Furthermore, the input noise of a single-balanced mixer is less than that of a double-balanced mixer [3]. This work also includes an illustration of a diode ring mixer circuit. In this case, the polarity of the RF voltage that appears at the IF output changes on a regular basis in response to changes in the local oscillator frequency. This also aids in frequency conversion, which is aided by the combined influence of the RF signal and the LO waveform [3].

h) Power consumption

Power consumption is a very potential issue in mixer design. Considering Power status, Power amplifiers (PAs) will have a very prominent role in 5G. For Wireless Communication System (WCS), transmitter has been very important element, with Power Amplifiers being the important RF part. Power consumption is a key concern in signal generators, frequency converters, signal processors and modulators through the use of Power Amplifiers. Phase tuning works to relax the design element of phase shifter along with digital baseband to reduce the power consumption [12, 13].

Massive Machine Type Communications (mMTC) can lead to high cost as well as high power consumption, which can be tackled using compressive sensing technique [14]. Besides, for 5G implementation, considering base station (BS) end of more bands in mmWave Bands may lead to increase complexity as well as increasing cost and power consumption [14]. High frequency multiplier can help in trade-offs between locking range and power consumption [14]. In power IC's design, if it is considered to minimize power consumption as well as heat dissipation, more importance need to be given towards high-frequency design. It also reduces volume of magnetic energy storage and corresponding system conversion loss [8, 15].

6.3 Diode Mixer

a) Single diode mixer

The theory of diode mixers is essential when designing a mixer for any other application. To begin, some assumptions have been made in our analysis. The following are the assumptions: (1) The RF signal should be smaller than the LO signal, which is also true in general. (2) The diode mixer should be stable under both large- and small-signal conditions.

The operating frequency of a diode-mixer is intended to be sensitive. It must also be sensitive to other frequencies, such as the well-known image frequency, which is present on the opposite side of the LO frequency from the RF. The problem is that undesired LO harmonics exist on both sides of the LO frequency, which is referred to as the sidebands. When a small-signal voltage is applied to the diode mixer, currents and voltages are generated in the junction at all of these frequencies, as well as sideband frequencies. These are known as small-signal mixing frequencies, and they can be written as n = 0 + n p, where n is the RF frequency, p is the LO frequency, 0 is the output or intermediate frequency, and n = ... -3, -2, -1, 0, 1, 2, 3. Without selecting any of the excitation frequencies, the diode's current, voltage, or both exist at each mixing frequency.

The diode device can be short-circuited to remove the voltage at any frequency, but other frequencies continue to occur. On the other hand, by leaving the diode open-circuit, the current components of the frequencies can be removed. The voltage, however, remains.

In theory, an infinite number of harmonics can exist. However, in practical applications, n is restricted to a certain maximum number in order to obtain the results of simulation studies.

b) Single balanced mixer

A single-ended mixer can be designed using a MOSFET with a load resistance. This device accepts a single-ended RF signal and a single-ended LO signal. The RF input has been applied to the source/drain contacts, and LO is connected to the MOSFET device's gate contact. Because the LO signal is generally unavailable during one-half of the LO cycle, the RF input signal is rejected during that time. That is why the information is lost during this process. An efficient topology for a differential LO input that commutates the RF input during each half of the LO cycle and results in differential output can be designed and implemented. Meanwhile, the LO port is balanced, but the RF input is still single-ended, so this topology is known as a "single-balanced mixer" topology. The gain of the single-balance mixer topology is two times that of the single-ended mixer. In the case of a single-balanced configuration, both passive and active mixers must be present. However, single-balanced topology is rarely used in modern transceivers. The most common problem with a single-balanced mixer is LO leakage caused by capacitance losses in a MOSFET device. The leakage of LO causes LO-to-IF feedthrough, which degrades the required or desired signal. The solution to the problem is described in the following section.

c) Diode ring mixer

Double-balanced mixers can perform LO noise cancellation, spurious signal rejection, and isolation between all ports and their intermodulation products. A higher LO drive was required for a doubly balanced diode mixer. The reason for the increased LO drive is that there are more diodes and their conversion loss is lower than in singly balanced designs. The operation of a diode ring mixer is very simple if the diode is viewed as a switch driven by the LO waveform. The diode ring mixer is made up of two center-tap transformers and four diodes arranged in a ring shape. Schottky diodes are used for better performance because of their low resistance and high-frequency response. The LO signal is applied to one transformer's primary winding. However, the secondary winding's center-tap is grounded at the same time. The RF signal is applied to the primary winding of the other transformer, and the IF output is taken from the center-tap of the secondary winding.

If you understand how a center-tapped transformer works, you can easily understand how a ring diode mixer works.

A center-tapped transformer with ideal and identical windings and configured to match all ports. As a result, the resistances at ports 3 and 4 are half of those at ports 1 and 2. When port 4 is excited, the current and voltage that appear across each port are as follows: same voltage appears across primary and secondary windings, if the current at port 4 is I, then the current at ports 1 and 2 is I/2, and the current at port 3 is zero to comply with power conservation. In most cases, the center tap is grounded, so no current flows. The center-tapped transformer, also known as a transformer hybrid, has a very broad bandwidth that is limited by winding capacitance and inductance. In practice, the center tap is usually grounded, so no current flows. The center-tapped transformer, also known as a transformer hybrid, has a very broad bandwidth. In high frequency design, this is limited by winding capacitance and inductance. When the identical load is applied to the transformers' center taps, they become virtual ground. It is linked to the two secondaries. As a result, the points where the secondary is virtual ground is also connected to the diodes. A and B are fictitious points. Points A and B are virtual grounds for the LO input, while points C and D are virtual grounds for the RF input.

6.4 Transistor Mixer

Transistor based mixer circuits significantly contribute towards majority of mixer designs. Emerging transistor technology utilizing Gallium

Nitride (GaN) materials has s prominent role in design of transistor mixer circuits for 5G applications. GaN based High Electron Mobility Transistor (HEMT) on Silicon carbide has led to open many options for considering high-temperature circuits. High temperature circuits are very vital in applications such as sensing, aerospace, automotive, jet-engine etc. Besides, for oil and gas industry, as it is drilled to extreme depths, higher temperature is most common. Bringing 5G into picture for these applications would require such transistor mixers based on high-temperature circuits. Higher bandgap increases runaway temperature for such GaN based transistors with higher thermal conductivity. Higher breakdown voltage with higher transition frequency leads to GaN HEMTs on SiC being extremely popular for high power RF applications [16].

There may be Resistive mixers which are based on field-effect transistors with great noise performance with low powered Local Oscillator, with high linearity and minimal DC power dissipation. These FET based passive mixers are having much better stability than active mixers. It shows a general resistive mixer topology. RF choles used in the transistor provide DC paths and helps transistor to act as variable resistor [16].

a) FET mixer

Field-Effect Transistor based mixer is one of the important design considerations in mixers. Rostomyan *et al.* has shown a FET based CMOS power amplifier with peak power added efficiency to be around 32.4 % [17]. Thirunavukkarasu *et al.* have considered balanced mixer designs which perform superior to single ended mixers in context of handling harmonics. These can be created using active devices such as FET or passive devices like diodes and capacitors. They have compared the performance of active as well as passive mixers for sub 6 GHz band frequency for 5G applications such as digital broadcasting [8].

Passive mixers which can be considered as folded version of active mixers, are relevant where having supply voltage is a limitation with high linearity demand. In this design, DC current will not be flowing through mixing FETs. They are further classified as a) Voltage-mode passive mixers and b) current-mode passive mixers [3]. Passive mixers generally are not affected by flicker noise as they commutate ac signal currents for DC of switches being equal to zero [3].

b) MOSFET Gilbert cell

Gilbert cell topology has been a very important design in context of mixers. Reja *et al.* has illustrated a novel CMOS wideband mixer where low voltage operation is assisted by switching transistors implemented in Gilbert Cell multiplier. Impedance matching condition has been successfully realized with this mixer considered to be suitable for Ultra-wideband applications [18]. Chang *et al.* have put forward a Gilbert cell topology, fabricated in 90-nm CMOS. It can achieve both up as well as down conversion through adjusting gate bias of Gilbert Switch [19].

Lal *et al.* illustrate another balanced mixer design of RF integrated circuit with single-balanced and double-balanced mixer. Double balanced mixer suffers from isolation issues. In this study, double balanced mixer has been used in Gilbert Cell architecture, with linearity improvement [20].

Mixers may play an important role in RF receiver structure with Gilbert cell providing high conversion gain. Kavivarman *et al.* have put forward gilbert mixer using resistive current reuse which improves conversion as well as forward gain to reduce noise figure. 5G systems would be supporting correspondence requirement of machine to machine or machine to human applications which require mixers, power amplifiers, transceivers etc. [21].

Gilbert cell which utilizes mixer core, may yield to high conversion gain as well as good port-to-port isolation [8]. In the work of Lianming *et al.* illustrates an up-converter mixer which consist of a Gilbert cell along with LO and RF buffer. RF buffer helps in boosting mixer output power. Transformer-based impedance matching network is further used to achieve inter-stage matching between Gilbert Cell, RF buffer as well as LO buffer [11]. It is made up of three components: a Gilbert cell, a LO, and an RF buffer. In order to realize the sliding-IF architecture, the LO frequency is set to 48 GHz [23]. The RF buffer was used to boost the mixer output power. We used a transformer-based impedance-matching network in this design to achieve inter-stage matching between the Gilbert cell, the RF buffer, and the LO buffer [11].

c) GaAs single gate switch

GaAs based technology can be used to help serve mmWave communication requirements. Millimeterwave *et al.* have demonstrated GaAs based components to serve requirements of mmWave communication as well as for sensor systems in range of 80 GHz [22]. Further, Lv *et al.* illustrates a work on Ka/Q dual-band Doherty power amplifier implemented with 0.1 μm Gallium Arsenide process. A novel method of designing mixers has been proposed where simple transmission lines have been used to help with phase-shift ranges in dual bands. To realize gain of Doherty Power amplifiers, reversed and uneven power splitter have been adopted to send more power [23]. Besides, for the limitations of GaAs based FET or corresponding switches, we have alternative options of GaN based architecture.

High power devices' modelling would require dynamic DC and RF techniques [24]. Besides, wide bandgap of GaN help in yielding superior properties, with much better performance than Si or GaAs devices. Monolithic power integration methods can be utilized to exploit potential of such devices [15]. Millimeter-wave (MMW) frequency bands which have been allocated for 5G wireless communication systems, would involve significant path-loss and poor transistor-based device performance. It needs to be effectively managed in upcoming technology versions [25].

6.5 Conclusion

In this chapter, several aspects of mixer designs have been considered. Several specifications of mixers were considered. Further, several examples have been considered for the illustration. Diode-mixers as well as several features of transistor mixers including their various forms were discussed in detail with detailed analysis of works from several research groups. Mixers are very vital component for the 5G communication architecture. Their integration into 5G communication architecture and further implementation along with other components will be very much crucial for future of 5G.

Acknowledgement

The authors are thankful to Department of Electrical Engineering, National Institute of Technology, Patna for providing resources to write this book chapter. And further they would like to acknowledge National Institute of Technology Patna and National Institute of Technology Srinagar respectively for providing healthy research environment.

References

1. Lin, W., Chen, J., Zhu, Y., Kong, F., Xu, J., Kuai, L., Yu, C., Yan, P., Hong, W., Wide band compact RF receiver for millimeter wave 5G mobile communication. *2016 IEEE International Conference on Ubiquitous Wireless Broadband, ICUWB 2016*, Institute of Electrical and Electronics Engineers Inc., 2016.
2. Al-Falahy, N. and Alani, O.Y.K., Millimetre wave frequency band as a candidate spectrum for 5G network architecture: A survey. *Phys. Commun.*, 32, 120–144, 2019.

3. Bashir, B., *Design of Integrated Mixer for 5G Radio Transceiver*, Thesis, Aalto University School of Electrical Engineering, 2019.
4. Kim, H.T., Park, B.S., Song, S.S., Moon, T.S., Kim, S.H., Kim, J.M., Chang, J.Y., Ho, Y.C., A 28-GHz CMOS direct conversion transceiver with antenna array for 5G cellular system. *IEEE J. Solid-State Circuits*, 53, 1245–1259, 2018.
5. Rollmann, G., Schmid, V., Mekhaiel, M., Knoll, P.M., Short range radar system for automotive applications, in: *Advanced Microsystems for Automotive Applications 2003*, pp. 215–221, Springer, Berlin Heidelberg, 2007.
6. Kao, J.C., Meng, C., Wei, H.J., Huang, G.W., 60-GHz SiGe BiCMOS dual-conversion down-converter: Schottky diode RF mixer and analog Gilbert IF mixer with microwave quadrature generator. *SiRF 2016-2016 IEEE 16th Topical Meeting on Silicon Monolithic Integrated Circuits in RF Systems*, Institute of Electrical and Electronics Engineers Inc., pp. 54–56, 2016.
7. Chen, Z., Liu, Z., Jiang, Z., Liu, P., Liu, H., Wu, Y., Zhao, C., Kang, K., A 27.5-43.5 GHz high linearity up-conversion CMOS mixer for 5G communication. *2017 IEEE Electrical Design of Advanced Packaging and Systems Symposium, EDAPS 2017*, January 2018, Institute of Electrical and Electronics Engineers Inc., pp. 1–3, 2018.
8. Thirunavukkarasu, G., A comparative study of active and passive mixers at sub 6 GHz for broadcasting applications. *Int. J. Emerg. Trends Eng. Dev.*, 8, 2913–2918, 2020.
9. Ding, Z., Zhang, D., Ma, C., Broadband antenna design with integrated CB-CPW and parasitic patch structure for WLAN, RFID, Wimax, and 5G applications. *IEEE Access*, 8, 42877–42883, 2020.
10. Gladson, S.C., Bhaskar, M., Praveen, R., Sudharsan, S., A 261-mW ultra-low power RF mixer with 26-dBm IIP3 using complementary pre-distortion technique for IEEE 802.15.4 applications. *AEU-Int. J. Electron. Commun.*, 107, 70–82, 2019.
11. Lianming, L. I, Xiaokang, N., Yuan, C., Linhui, C., Tao, Z., Depeng, C., The path to 5G: mmWave aspects. *J. Commun. Inf. Netw.*, 1, 1–18, 2016.
12. Qi, T. and He, S., Power up potential power amplifier technologies for 5G applications. *IEEE Microw. Mag.*, 20, 6, 89–101, 2019.
13. Hu, S., Wang, F., Wang, H., A 28-/37-/39-GHz linear doherty power amplifier in silicon for 5G applications. *IEEE JSSC*, 54, 1586–1599, 2019.
14. Huo, Y. and Dong, X., Enabling multi-functional 5G and beyond user equipment: A survey and tutorial. *IEEE Access*, 7, 116975–117008, 2019.
15. Sun, R., Lai, J., Chen, W., GaN power integration for high frequency and high efficiency power applications: A review. *IEEE Access, Special Section on Emerging Technologies for Energy Internet*, 8, 15529–15542, 2020.
16. Salem, J.M. and Pour, F.L., A high-temperature model for GaN-HEMT transistors and its application to resistive mixer design. *IEEE Trans. Circuits Syst.*, 68, 581–591, 2021.
17. Rostomyan, N., Jayamon, J.A., Asbeck, P., 15 GHz 25 dBm multigate-cell stacked CMOS power amplifier with 32 % PAE and ≥ 30 dB gain for 5G

applications, *Proceedings of the 11th European Microwave Integrated Circuits Conference,* 15, pp. 265–268, 2016.
18. Reja, M., Moez, K., Filanovsky, I., *A Novel 0. 6V CMOS Folded Gilbert-Cell Mixer for UWB Applications*, pp. 169–172, 2008.
19. Chang, Y. and Lin, K., A 28-GHz bidirectional active Gilbert-cell mixer in 90-nm CMOS. *IEEE Microw. Wirel. Compon. Lett.*, 31, 473–476, 2021.
20. Lai, W. and Huang, J., *A 3~5GHz Broadband High Linearity CMOS Gilbert-Type Mixer and Low Power LNA for UWB transceiver on Intelligent Control.*
21. Kavivarman, J., Partibane, B., Vaithianathan, V., *Applications Using 45nm CMOS Technology*, pp. 33–39, 2020.
22. Daembkes, H., Adelseck, B., Schmidt, L.P., Schroth, J., Daimler-Benz, A.G., Research Center Ulm, *In IEEE NTC'95, the Microwave Systems Conference: Conference Proceedings (p. 83). Institute of Electrical & Electronics Engineers (IEEE),* Daimler-Benz Aerospace, Ulm, Germany, 1995.
23. Lv, G., Chen, W., Chen, X., Ghannouchi, F.M., A compact Ka/Q dual-band GaAs MMIC doherty power amplifier with simplified offset lines for 5G applications. *IEEE Transactions on Microwave Theory and Techniques*, 67, 3110–3121, 2019.
24. Yuk, K., Branner, G.R., Cui, C., *Future Directions for GaN in 5G and Satellite Communications*, pp. 803–806, 2017.
25. Park, H., Kang, D., Lee, J., Minn, D., Aoki, Y., Kim, K., Lee, S., Lee, D., Kim, S., Kim, J., Lee, W., Kim, C., Park, S., Park, J., Suh, B., Jang, J., Kim, M., Min, K., Jeon, S., Ryu, A., Kim, Y., Lee, J.H., Son, J., Yang, S., *Millimeter-Wave Band CMOS RF Phased-Array Transceiver IC Designs for 5G Applications*, pp. 355–358, 2020.

7

RF LC VCOs Designing

M. Sankush Krishna[1]*, Madhuraj Kumar[1], Neelesh Pratap Singh[1,2] and Anjan Kumar[1,3]

[1]*Department of Electronics and Communication Engineering, National Institute of Technology, Patna, Bihar, India*
[2]*Quality Assurance & Inspection Circle, Bharat Sanchar Nigam Limited, Jabalpur, India*
[3]*Department of Electronics and Communication Engineering, GLA Mathura, U.P., India*

Abstract

One major part of the primary components of the transceiver is the RF voltage regulated oscillator (VCO). Inductor (L)-capacitor (C) VCOs and ring VCOs are the two primary types of VCOs. Ring oscillators have a large tuning range and a little area occupation. For most wireless applications, however, they frequently result in intolerable phase noise. Because of the current necessary to maintain the tank oscillation, using an LC oscillator improves phase noise while wasting a lot of power. As they produce little jitter or phase noise when contemplated in the frequency domain, LC oscillators are preferred in the clock conversion function of optical networking systems. An LC-phase oscillator's noise is correlated to the inductor's quality factor. The smaller the phase noise, the greater the quality factor, but the larger the inductor size, the higher the circuit cost. The noise characteristics of these RF VCOs are studied in order to get the lowest possible consumption. In this literature, LC oscillators are considered because they are the multifarious interesting for high-frequencies. Up to very high frequencies, LC oscillators can be utilized to provide output signals with a foreseeable frequency of oscillation and low phase noise.

***Keywords*:** LC VCO, Q-factor, phase noise

**Corresponding author*: sankushnov2011@gmail.com

Sangeeta Singh, Rajeev Kumar Arya, B.C. Sahana and Ajay Kumar Vyas (eds.) RF Circuits For 5G Applications: Designing with mmWave Circuitry, (123–136) © 2023 Scrivener Publishing LLC

7.1 Introduction

The accelerated growth in network communication systems has demanded better transceiver systems. Oscillators are the common and yet important functional blocks in the communication systems. In the communication system ICs, integrated voltage regulated oscillators have been the subject of a lot of research in the last two decades. Generally, ring oscillator (RO) VCOs and LC VCOs are two types of VCOs used in the communication systems. RO VCOs have a number of delay cells designed in a way to sustain the Barkhausen criteria. Since they do not have inductors, they have a compact chip area and a high tuning range. In spite of this precedence, the phase noise of the ring oscillators is extremely high. Hence, the researchers have started to opt for LC VCOs. The LC VCOs have resonator LC tank with inductor and capacitor. The privation from physical components can be defined by a resistance which can be compensated a negative resistance from the active devices in the circuit. The center frequency can be changed by changing the capacitance. A reasonable tradeoff is chosen amidst the frequency tuning range and the capacitance value as at higher frequencies the varactor quality factor degrades thereby inducing phase noise. Main shortcoming of the LC VCO is the usage of an inductor which consumes larger chip area. Several factors, including phase noise, oscillation frequency, tuning range, figure of merit (FOM), and power consumption are taken into account when evaluating the performance of VCOs. Various cross coupled topologies of LC VCOs are compared and tabulated by Moghavvemi *et al.* [1].

7.1.1 Basic VCO Models

Basically different topologies of the VCOs are used such as quadrature VCOs, colpitts LC VCOs, and cross-coupled LC-VCOs. Quadrature LC VCO (QVCO) provides local oscillation quadrature signal for signal up and down-conversion. The quadrature signals can be generated by a divide-by-two circuit after a differential oscillator, polyphase filter or the coupling between the differential oscillators. QVCO comprises two VCO cores coupled with each other providing an advantage of lower power consumption. Colpitts oscillators have superior cyclostationary noise properties which enables them to achieve significantly low phase noise. In spite of these advantages, the integrated circuits in these days hardly use colpitts oscillator topologies. They demand higher gain for definitive start-up and their single ended character projects it more likely to modify parameters and common-mode noises like substrate and supply noise. The most special design for the VCO topology is the cross coupled design. They have advantages such as ease of design, better start-up conditions, and differential operation. When the oscillator is susceptible to perturbations, noise from active devices is generated, which degrades the phase noise. A few designs have been presented in the past to minimize noise from cross coupled transistors [2–4].

7.1.2 Phase Noise

The fundamental difficulty in designing fully integrated LC VCOs is achieving low phase noise though keeping power consumption to a minimum. The poor-quality factor of the inductor could lead to the phase noises from the LC tank. In a VCO, the tail transistor may be the most significant inception of phase noise. The noise contributed by the transistors can be categorized into two different ways. One is the high frequency up-converted flicker noise and the other is the low frequency phase noise. A hard switching oscillator converts the high frequency up-converted flicker noise from the tail transistor into phase noise at double the oscillation frequency. The other source of the phase noise at low frequency is introduced by the tail transistor during various mechanisms such as AM-to-PM transfiguration in the non-uniformity of the varactor [5], modulation of the bias point [6], modulation of tail capacitance and Groskowski effect [7]. Different models are proposed for the LC VCOs [8]. The differentially cross coupled LC VCO consumes less power and provides best FOM. The complementary cross coupled LC VCOs have higher tuning range and compensate the active power loss. The cross coupled pseudo resistance LC-tank design achieves low power consumption without reduction in phase noise. It consumes low current and provides best FOM. Tom Dreschel *et al.* [9] reported ultra-wide band SiGe LC VCO with 80 % continual tuning range. A cross coupled oscillator is proposed with an output stage such that the output is matched to 50 W. The output stage and the LC tank are decoupled to avoid the further addition of the capacitances to the tank using the common collector stage using a BJT transistor. A low phase noise of about -110 dBc/Hz is noticed at 1 MHz offset frequency. A high steady tuning range of about 80 % is obtained from 6.5 to 15.1 GHz with a low power consumption of about 19 mW. Xi *et al.* [10] proposed a QVCO with dual tail current biasing to attain low phase noise and high accuracy. Dual tail current method inserts a phase shift due to which, the total quality of QVCO and amplitude of output signals are augmented thereby improving the phase noise. Also, DC biasing voltages of the tail transistors makes the phase shift reconfigurable to process variations. The VCO attains a low phase noise of about -129 dBc/HZ at 1 MHz offset value. Excellent FOM of about 184.1 dBc/Hz is achieved. Koukab *et al.* [11] reported LC VCO with two-fold g_m to obtain enhanced RF oscillation and attenuated low frequency noise. An amended self-biased prescaler is singularly attuned to this topology to avoid noisy and power consuming buffers and DC bias shifters was reported. The design is proposed in a way to have high g_m at higher frequencies to boost start-up oscillation states with optimized phase noise. Low g_m at low frequencies reduces the switching transistors noise and reducing the phase noise at low frequencies. Power supply rejection ratio (PSRR) is also improved. The fabricated circuit achieved the highest FOM possible 189.2 dBc/Hz/mW at 1 MHz offset, proving the good phase noise/power consumption barter. Zou *et al.* [12] reported Ku band VCO using dual LC tank with transformer-based feedback using the BiCMOS process. A triple coil transformer is utilized to come up with robust coupling between the two LC tanks and the transformer-based feedback. The transformer coil's parameters, which are critical in ascertaining phase noise, oscillation, and output signal swing, are improved. The strong coupling between the two LC tank and the transformer feedback is found to improve the quality factor. A low phase noise of about -117.4 dBc/Hz is achieved at 1 MHz offset from the 12.64 GHz oscillation frequency. The power consumption of the VCO core is the 4.3 mW. Choi *et al.* [13] reported source injection parallel coupled quadrature VCO (SIPC-QVCO). Unlike parallel QVCO, the drain connections of the coupling transistors are directed to the supply voltage. The 1/f noise introduced by the coupling transistors is reduced by diverting the drain connections to an AC ground. For the SIPC-QVCO design, the phase noise measured at 1.5 GHz is noticed to have greater than 10 dB enhancement compared to that of the convention P-QVCO over the 1 MHz offset frequency while at the same amount of power dissipation. The current noise from the cross coupled transistor pairs to the LC tank is bring down by the complementary cross coupled pair topology implemented in the VCO circuit, a differential boot strapped inductor which enacts a high-Q active inductor specified by a low noise improvement.

7.1.3 Flicker Noise

Additionally to the white thermal noise, MOS transistors also introduce the flicker noise which has the power spectral density inversely correlated to frequency. This noise up-converts to higher frequencies in VCOs. Different researchers proposed various methods to reduce the flicker noise. Boon *et al.* [14] reported a method to reduce the noise of CMOS LC oscillator.

The 1/f noise can be lessened by the switching gate and the switching gate frequency is inversely correlated to the flicker noise. A fully integrated CMOS oscillator accompanied by a tank quality factor of about 9 achieves a FOM of roughly 193 dB. The proposed topology is contrasted to the conventional topologies such as fixed biasing tail transistor and without tail transistor topologies. The main advantage of the proposed method is that the only source of flicker noise is the cross coupled transistors undergo with a lower flicker noise owing to switched biasing. Fong *et al.* [15] reported wide-band VCO for common mode noise rejection with differentially tuned accumulation MOS varactors using CMOS SOI technology. The MOS varactors provide a tuning range of about 40 % within 0-1 V biasing window. It also reduces the result of high varactor sensitivity by rejection of common mode noise. For differential LC VCOs, the low frequency noises like flicker noise can be regarded as common mode noise and the common mode noise can be suppressed from being up-converted to carrier frequency by using differentially tuned varactors. The phase noise measured is around -121.67 dBc/Hz over 1 MHz offset frequency at 3.8 GHz and -111.67 dBc/Hz at 5.7 GHz frequency. The dissipated power is around 2.3-2.7 mW, based on center frequency and the buffered output power is 9 dBm. The VCO can be operated at very low voltage and low power because of the noise rejection. It dissipated a power of about 0.8 mW at 5.5 GHz while the voltage is at 0.75 V. Hoshino *et al.* [16] reported differential LC oscillator, that nulls the flicker noise up-conversion. An array of design techniques is used in the VCO circuit to vanquish the numerous techniques that are accountable for the flicker noise up-conversion. It is also designed for lowest Leeson's factor that causes phase noise with less thermal induction. The approach is robust and based on understanding the physical mechanisms of the phase noise specifically about the flicker noise up-conversion. It has accredited wide tuning range of the CMOS VCO to linkup the demands for close-in phase noise. The prototype that is intended to use in dual conversion receiver dissipates a power of about 2.5 mW and gives a FOM of 195 dB. Without noise filter the FOM drops to 187 dB. Muer *et al.* [17] reported a fully integrated 2 GHz very low-phase noise LC tank VCO set with flicker noise up-conversion minimization implemented in three metal layer 0.65 μm BiCMOS process using only MOS active devices. The phase noise measured is as low as -125.1 dBc/Hz at 600 kHz and -138 dBc/Hz at 3 MHz offset frequency due to the use of the planar integrated inductors. A flicker noise up-conversion factor is determined which minimizes the flicker noise up-conversion to 1/3rd phase noise. The flicker noise up-conversion factor is obtained from the circuit simulation. The proposed method demonstrates the correlation betwixt the flicker noise up-conversion and the overdrive level of the MOS cross coupled pair of the oscillator. The circuit balancing techniques are also developed to reduce it even more up-conversion of the flicker noise. The $1/3^{rd}$ phase-noise corner is kept to less than 15 kHz.

7.1.4 Distributed Oscillators

LC VCOs provide better phase noise performance but it is not possible to achieve all specifications of VCO simultaneously. Particulalry, at the self-resonance frequency of the on-chip inductors and resonance frequency of the MOS transistors. The inductor losses, loading from the next stages such as buffer stages, and parasitic capacitances of the transistors do not scale at the same rate. In addition to operate at elevated frequencies, minimum LC tank product is desired which demands smaller L. Using smaller L further increases the power dissipation for the given oscillation frequency as well as other rigorous limits on the ability to tune such LC VCOs [18]. Due to these restrictions, alternate techniques such as distributed oscillators must be pursued. Distributed amplifiers often offer gain over tens of gigabits of bandwidth, so an electronically controllable oscillator with a similar tuning range would be interesting. Kleveland *et al.* [19] reported 0.18 μm CMOS 17 GHz distributed oscillator using the forward gain operation mode with off-chip termination and without any tuning capability. White *et al.* [20] reported phase noise in distributed oscillators. The effective capacitance, which is equal to the total capacitance in the transmission lines, can be used to determine phase noise in distributed oscillators. To improve the overall Q of a distributed oscillator, the power loss at the terminations must be reduced. Li *et al.* [21] reported distributed dual band LC oscillator for low phase noise with switched resonator design implementation. The mode carrying switches do not carry current and hence, the Q factor is unaffected thereby leading to low phase noise. It achieves a FOM of 194.5 dB at the 4.9 GHz band and 193.0 dB at the 6.6 GHz band, which is the same as the reference standalone LC oscillator.

Challenges at mmWave Frequencies

In the operation of fully integrated transceiver layouts, VCOs are essential. Quadrature local oscillator (LO) signals are required for frequency conversion in the majority of applications. The creation of such quadrature phase outputs with minimal power consumption is a major problem, especially in portable applications. This also applies to upcoming mmWave applications, where a broader tuning range is desired while keeping power consumption and phase noise low. We look at many design strategies for generating a high-performance wireless receiver building block in this paper, with a particular focus on LC-based VCOs. One of the difficulties for mmWave VCOs is that the quality factor decreases as the operating frequency rises. Due to this the phase noise degrades in thermal region and the output power decreases. One more concerning factor is the flicker noise which gets worse for mmWave operating VCOs. For small scale VCO CMOS technologies it gets even worse. Another major concern for designing the VCOs is the parasitic noises. The parasitic capacitances might degrade the tuning range. This is in particular serious for the digitally controlled oscillators at mmWave range. As the operating frequency is increased, the smaller L is desired which further increases the power dissipation for a given oscillation frequency.

7.2 Tuning Extension Techniques

In this section, we talked about how the knowledge we've gained so far has been in relation to the architecture of tuning circuits for mmWave oscillations.

7.2.1 Varactor

Varactor is also called varicap diode. It is used to provide voltage controlled capacitance in RF circuit design. It is used in that circuit design where capacitance can be controlled by voltage. To design a low-GHz range frequency tuning circuit accumulation mode MOS(A-MOS) varactor is generally employed. A-MOS varactor are generally implemented alongside switched capacitor to extend the TR without interrupting the AM-PM noise to phase noise conversion method, reduce the actuation level in analog phase locked loop by diminishing the VCO gain K_{VCO}, it also ensure stability. When we use an A-MOS varactor of channel length 200 nm realize in a 28 nm CMOS process technology, it shows C_{MAX}/C_{MIN} of 3.9:1 with quality factor range varied from 9.1 to 2.6 at frequency 80 GHz because quality factor of capacitance depends on frequency. Varactor diode is widely used in voltage controlled oscillator (VCO), RF filters, frequency and phase modulators.

$$Q_C = \frac{1}{2\pi f R_c}$$

7.2.2 Switched Capacitors

A switched capacitor is an electronics device. It is used in a filter. Its working is based on movement of charges. When switch is open charge moves into the capacitor and when switches are closed charge depletes from capacitor. Filters which are implemented with switched capacitor are known as "switched capacitor filters", and depends on the ratio between capacitances of used capacitors, this behavior of switched capacitor makes it suitable to use within integrated circuits (IC), since in IC specified capacitors and resistors making is more cost effective. For implementing a SC circuits, metal oxide semiconductor technology are used with MOS capacitors and MOSFET switches, and all these are commonly fabricated using CMOS process. The quality factor of a switched capacitor diminishes with each technological node and with relation to the frequency of operations, as we know. When we used deep scaled CMOS for capacitor, figure of merit of switches has a proclivity to saturate and hence it does not perform best for mmWave. In general switched capacitors are used in digital to analog converter, analog to digital converter, pulse code modulation etc. For a differential switched capacitor bank owing to parasitic of a real switch the best attainable C_{MAX} to C_{MIN} ratio is expressed as

$$\frac{C_{MAX}}{C_{Min}} = \frac{C_{MOM}}{2C_{SW,OFF}}$$

Switched capacitor bank achieves minimum quality factor in on state and it is expressed as

$$Q_{Min} = \frac{2}{\omega C_{MOM}(2R_{MOM} + R_{SW,ON})}$$

Above two equations shows the barter between minimum quality factor and tuning range. When we use -Gm model then the minimum and maximum frequency in off and on state is

$$f_{min} = \frac{1}{2\pi\sqrt{C_{PAR} + C}}$$

$$f_{MAX} = \frac{1}{2\pi\sqrt{L\left(C_{PAR} + \frac{CC_{SW,OFF}}{C + C_{SW,OFF}}\right)}}$$

When C_{PAR} increases f_{MAX} decreases and it limits the maximum achievable f_{MAX}. Hence, we focus on another alternative tuning technique which is appropriate for tuning extension at mmWave.

7.2.3 Switched Inductors

With constant scaling of the integrated circuit, the quality factor of inductors does not degrade much, although it does improve with frequency, hence many researchers are working on different techniques to switch inductors and successfully proposed at mmWave. Mamei *et al.* proposed a LC-tank oscillator in which they have used a switch in series to the capacitor and inductor [22]. When they simulate the circuit then they found that when switch is ON the minimum resonance frequency is

$$f_{min} = \frac{1}{2\pi\sqrt{C_T + C_{fix}}}$$

And when switch is in OFF condition then the maximum resonance frequency is

$$f_{MAX} = \frac{1}{2\pi\sqrt{L_T\left(\frac{(C_{fix}+C_T)C_{SW,OFF}}{C_{fix}+C_T+C_{SW,OFF}}\right)}}$$

In this case f_{MAX} is not controlled by parametric capacitance, further capacitance in switch off condition is allowed to be large. When we improve resistance of switched capacitor in on state it further improve Q_{min}. For further improvement Demirkan *et al.* designed a wide tuning-range CMOS VCOs by using switched coupled inductors [23]. They have found that when switched on then equivalent inductance is

$$L_{eq,on} = L_1 - \frac{\omega^2 M^2 L_2}{(R_2 + R_{on})^2 + \omega^2 L_2^2} \approx L_1 - \frac{M^2}{L_2}$$

Where L_1 and L_2 are inductance of primary and secondary coil whereas M is correlative inductance between them and R_{on} is the resistance of the switch in on state and ω is the angular frequency, R_2 is ohmic loss of secondary coil. Whereas when switch is in OFF state, R_2 is sufficiently small and $\omega < \frac{1}{\sqrt{L_2 C_{OFF}}}$ then equivalent inductance is expressed as

$$L_{eq,OFF} = L_1 - \frac{\omega M^2}{\omega L_2 - (\omega C_{off})^{-1}}$$

7.2.4 Switched TLs

Another approach, based on the dispersed features of transmission lines (TLs), has been devised by certain researchers that are more effective at mmWave. TLs have the greatest Q factor, which changes when metal strips are appended. Properties of the TLs controlled by adding the switches and the emerged circuit act as like distributed capacitor bank. To design a 60 GHz CMOS amplifier LaRocca *et al.* used differential TLs design, to improve the quality factor and handle 40 mA bias current. At the last stage they have used a shorted stub of 20 μm wide lines and 10 μm gap. Used TLs transforms the real output load to inductive load, hence it compensate the output capacitance of the circuit design. To keep the low inductance we use the shorted stub of low characteristics impedance, it reduces the length of TLs [24]. They have also designed a digitally tuned 58-64 GHz digital controlled oscillator (DCO) in which a digital controlled differential transmission lines are embedded in 90 nm CMOS [25]. They have also design an embedded DiCAD linear phase shifter for 57-65 GHz DCO using differential transmission lines [25]. Wu *et al.* also used this technique to design a two millimeter-wave DCO that achieve both tuning range of 10% and fine frequency resolution of 1 MHz at the same time [26].

7.2.5 4th Order Tanks and Other Techniques

Some researchers used other tuning technique except switched capacitor, switched inductors, switched TLs and Varactor. As we discussed many tuning techniques in previous section, there are many favorable technology used for scaling as well as mmWave applications. Bevilacqua *et al.* used another popular technique, in which technique they have used 4th order tank circuit, it exploits two resonant frequencies. This work used transformer based resonator one port network and as a two-port network. Both configuration shows similar quality factor, it free from design parameters however when coil is loosely coupled then resonator enhanced the operation for high frequency. They designed a VCO based on these findings and traditional tuning techniques, range of VCO is 3.6-7.8 GHz [27].

Yin *et al.* design a magnetically tuned multimode CMOS VCO of tuning frequency range is 57.5-90.1 GHz. In this work to enhance the frequency tuning range they have used a switched-triple-shielded transformer, it changes the magnetic coupling coefficient between its primary and secondary coils. They have used 65-nm CMOS technology process to design the magnetically tuned VCO. Magnetically tuned VCO design using this technique measures a frequency range of 41.1% of its tuning range (57.5-90.1 GHz) when supply is 1.2 V, current consumption is 7 to 9 mA and chip area is 0.03 mm^2 [28].

However, when compared to the standard LC tank, it has been shown reportedly that high order resonators do not supply any fundamental PN increase. Vigilante *et al.* design an E-band VCO in 28-nm CMOS process technology. To get an accurate quadrature phase, two oscillators are connected using gate to drain transformers, and switched inductors are utilized for tuning extension [29]. Mazzanti *et al.* designed a transformer based one port and a two port resonator to investigate the phase noise (PN) or figure FOM and they found that one port transformer based resonator does not provide any benefit in comparison to simple LC tank design [30]. Li *et al.* design a Colpits LC VCO in 90 nm CMOS process technology. This technique enhances the transconductance and reducing the phase noise; this oscillator is ahead of the tank oscillator [31].

7.3 Conclusion

According to current market demands and industry trends, communication equipment will need to accommodate bigger bandwidths and an ever-increasing variety of frequency bands and protocols. Many components of such devices face interesting hurdles as a result of this. In RF transceivers, the VCO is among the most important component. The trade-offs in design are presented in this work. Low-power LC-VCOs are discussed as well as a design strategy for RF VCOs. In addition, phase noise has been looked into. It is feasible to establish whether an oscillator covering the frequency range for the target application is viable using the theory offered and some basic information about the existing IC technology.

Acknowledgement

The authors are indebted to the National Institute of Technology, Patna's Department of Electronics and Communication Engineering for providing resources to compose this book chapter.

References

1. Moghavvemi, M. and Attaran, A., Performance review of high-quality-factor, low-noise, and wideband radio-frequency LC-VCO for wireless communication [application notes]. *IEEE Microw. Mag.*, 12, 4, 130–146, 2011.
2. Tsai, M.-D., Cho, Y.-H., Wang, H., A 5-GHz low phase noise differential colpitts CMOS VCO. *IEEE Microw. Wirel. Compon. Lett.*, 15, 5, 327–329, 2005.
3. Chu, Y.K. and Chuang, H.-R., A fully integrated 5.8 ghz u-nii band 0.18-micro m CMOS VCO. *IEEE Microw. Wirel. Compon. Lett.*, 13, 7, 287–289, 2003.
4. Ho, C.-C., Kuo, C.-W., Hsiao, C.-C., Chan, Y.-J., A 2.4 GHz low phase noise VCO fabricated by 0.18/spl mu/m pmos technologies, in: *2003 International Symposium on VLSI Technology, Systems and Applications. Proceedings of Technical Papers. (IEEE Cat. No. 03TH8672)*, IEEE, pp. 144–146, 2003.
5. Samori, C., Lacaita, A.L., Zanchi, A., Levantino, S., Torrisi, F., Impact of indirect stability on phase noise performance of fully-integrated LC tuned VCOs, in: *Proceedings of the 25th European Solid-State Circuits Conference*, IEEE, pp. 202–205, 1999.
6. Samori, C., Levantino, S., Boccuzzi, V., A-94 dbc/Hz@ 100 KHz, fully-integrated, 5-GHz, CMOS VCO with 18% tuning range for bluetooth applications, in: *Proceedings of the IEEE 2001 Custom Integrated Circuits Conference (Cat. No. 01CH37169)*, IEEE, pp. 201–204, 2001.
7. Rael, J. and Abidi, A.A., Physical processes of phase noise in differential LC oscillators, in: *Proceedings of the IEEE 2000 Custom Integrated Circuits Conference (Cat. No. 00CH37044)*, IEEE, pp. 569–572, 2000.
8. Nandanwar, M. and Dandekar, D., High performance VCO for low power rf transmitter-a review, in: *National Conference on Innovative Paradigms in Engineering & Technology (NCIPET-2012)*, Citeseer, pp. 22–25, 2012.
9. Drechsel, T., Joram, N., Ellinger, F., 6.5 to 15.1 GHz ultra-wideband sige LC VCO with 80% continuous tuning range, in: *2017 European Conference on Circuit Theory and Design (ECCTD)*, IEEE, pp. 1–4, 2017.
10. Xi, N., Lin, F., Ye, T., A low phase noise, high phase accuracy quadrature LC-VCO with dual-tail current biasing to insert reconfigurable phase delay. *IEEE Trans. Circuits Syst. II Express Br.*, 67, 3, 450–454, 2019.
11. Koukab, A., LC-VCO design with dual-Gm boosted for RF oscillation and attenuated for lf noise. *IEEE Microw. Wirel. Compon. Lett.*, 20, 12, 675–677, 2010.
12. Zou, Q., Ma, K., Yeo, K.S., Lim, W.M., Design of a ku-band low- phase-noise VCO using the dual LC tanks. *IEEE Trans. Circuits Syst. II Express Br.*, 59, 5, 262–266, 2012.
13. Choi, H.C., Shin, S.B., Lee, S.-G., A low-phase noise LC-QVCO in CMOS technology. *IEEE Microw. Wirel. Compon. Lett.*, 14, 11, 540–542, 2004.

14. Boon, C.C., Do, M.A., Yeo, K.S., Ma, J., Zhang, X.L., RF CMOS low-phase-noise LC oscillator through memory reduction tail transistor. *IEEE Trans. Circuits Syst. II Express Br.*, 51, 2, 85–90, 2004.
15. Fong, N.H., Plouchart, J.-O., Zamdmer, N., Liu, D., Wagner, L.F., Plett, C., Tarr, N.G., 1-V 3.8-5.7-GHz wide-band VCO with differentially tuned accumulation MOS varactors for common-mode noise rejection in CMOS SOI technology. *IEEE Trans. Microw. Theory Tech.*, 51, 8, 1952–1959, 2003.
16. Hoshino, K., Hegazi, E., Rael, J., Abidi, A., 1.5 V, 1.7 ma 700 MHz CMOS LC oscillator with no upconverted flicker noise, in: *Proceedings of the 27th European Solid-State Circuits Conference*, IEEE, pp. 337–340, 2001.
17. De Muer, B., Borremans, M., Steyaert, M., Puma, G.L., A 2-ghz low-phase-noise integrated LC-VCO set with flicker-noise upconversion minimization. *IEEE J. Solid-State Circuits*, 35, 7, 1034–1038, 2000.
18. Ham, D., Design of integrated LC VCOs, in: *Trade-Offs in Analog Circuit Design,* Springer, 517–549, 2002.
19. Kleveland, B., Diaz, C., Vock, D., Madden, L., Lee, T., Wong, S., Monolithic CMOS distributed amplifier and oscillator, in: *1999 IEEE International Solid-State Circuits Conference. Digest of Technical Papers. ISSCC,* First Edition (Cat. No. 99CH36278), IEEE, pp. 70–71, 1999.
20. White, C. and Hajimiri, A., Phase noise in distributed oscillators. *Electron. Lett.*, 38, 23, 1453–1454, 2002.
21. Li, G. and Afshari, E., A distributed dual-band LC oscillator based on modes witching. *IEEE Trans. Microw. Theory Tech.*, 59, 1, 99–107, 2010.
22. Mammei, E., Monaco, E., Mazzanti, A., Svelto, F., A 33.6-to-46.2GHz 32nm CMOS VCO with 177.5dBc, Hz minimum noise FOM using inductor splitting for tuning extension, in: *2013 IEEE International Solid-State Circuits Conference Digest of Technical Papers*, San Francisco, CA, pp. 350–351, 2013.
23. Demirkan, M., Bruss, S.P., Spencer, R.R., Design of wide tuning-range CMOS VCOs using switched coupled-inductors. *IEEE J. Solid-State Circuits,* 43, 5, 1156–1163, 2008.
24. La Rocca, T., Liu, J.Y.C., Chang, M.C.F., 60 GHz CMOS amplifiers using transformer-coupling and artificial dielectric differential transmission lines for compact design. *IEEE J. Solid-State Circuits,* 44, 5, 1425–1435, 2009.
25. La Rocca, T., Liu, J., Wang, F., Chang, F., Embedded DiCAD linear phase shifter for 5765GHz reconfigurable direct frequency modulation in 90nm CMOS, in: *2009 IEEE Radio Frequency Integrated Circuits Symposium,* Boston, MA, pp. 219–222, 2009.
26. Wu, W., Long, J.R., Staszewski, R.B., High-resolution millimeter-wave digitally controlled oscillators with reconfigurable passive resonators. *IEEE J. Solid-State Circuits,* 48, 11, 2785–2794, 2013.
27. Bevilacqua, A., Pavan, F.P., Sandner, C., Gerosa, A., Neviani, A., Transformer-based dual-mode voltage-controlled oscillators. *IEEE Trans. Circuits Syst. II Express Br.*, 54, 4, 293–297, 2007.

28. Yin, J. and Luong, H.C., A 57.590.1-GHz magnetically tuned multimode CMOS VCO. *IEEE J. Solid-State Circuits*, 48, 8, 1851–1861, 2013.
29. Vigilante, M. and Reynaert, P., Analysis and design of an E-band transformer-coupled low-noise quadrature VCO in 28-nm CMOS. *IEEE Trans. Microw. Theory Tech.*, 64, 4, 1122–1132, 2016.
30. Mazzanti, A. and Bevilacqua, A., On the phase noise performance of transformer-based CMOS differential-pair harmonic oscillators. *IEEE Trans. Circuits Syst. I Regul. Pap.*, 62, 9, 2334–2341, 2015.
31. Li, L., Reynaert, P., Steyaert, M., A colpitts LC VCO with Miller-capacitance Gm enhancing and phase noise reduction techniques, in: *2011 Proceedings of the ESSCIRC (ESSCIRC)*, Helsinki, pp. 491–494, 2011.

8

RF Power Amplifiers

Anchal Tyagi[1], Rachit Patel[1*] and Krishna Pandey[2]

[1]*Department of Electronics and Communication Engineering, ABES Institute of Technology, Ghaziabad, India*
[2]*Department of Electronics and Communication Engineering, National Institute of Technology, Patna, Bihar, India*

Abstract
In this chapter, the basic designing and characteristics of the RF Power Amplifier (PA) are discussed in depth. RF Power Amplifier designs are specially considered for various 5G and higher frequency applications. Here various approaches used to achieve and to optimize the performance of the Power Amplifier. Both the designing, Bipolar Power Amplifier and CMOS Power Amplifier discussed here. This chapter covers the designing parameters for RF Power Amplifier, classification of Power Amplifiers, different designing approaches to design the Bipolar PA and CMOS PA for the application in 5G. Linearization principles also discussed at the end of the chapter.

Keywords: RF power amplifier, efficiency, bipolar PA design, CMOS PA design, cascode structure, linearization

8.1 Specification

In this section, the different important specifications are discussed. Power amplifiers are being commonly used in the telecommunication system for delivering some amount of power to the antenna. Telecommunication standard define certain specification according to their technology i.e. UMTS specification is 24dBm (output power).

**Corresponding author*: rachit05081gece@gmail.com

Sangeeta Singh, Rajeev Kumar Arya, B.C. Sahana and Ajay Kumar Vyas (eds.) RF Circuits For 5G Applications: Designing with mmWave Circuitry, (137–156) © 2023 Scrivener Publishing LLC

8.1.1 Efficiency

An RF front end always suffers some losses due to follower circuit which is compensated by the power amplifier. *Output efficiency* of the amplifier is defined as the ratio between RF output power versus the dc input power. It can be represented by

$$\zeta = \frac{P_{ro}}{P_{dc}} \tag{8.1}$$

Whereas Pro, and Pdc are RF output power and dc input power respectively. With taking into the consideration input power delivered, the new measure of efficiency can be calculated named as power added efficiency (PAE).

$$PAE = \frac{P_o - P_{in}}{P_{dc}} = \zeta\left(1 - \frac{1}{G}\right) \tag{8.2}$$

Here, G represents the power gain and Pin is input power delivered. Normally the efficiencies can vary in the range of 35–55% [1, 2].

8.1.2 Generic Amplifier Classes

Conduction angle, Input Signal, and Load impedance are the parameters which characterize the behavior of an amplifier. On the basis of efficiency, it can be categorized into the six classes namely Class A, Class B, Class C, Class D, Class E, and Class F.

- **Class A:** It works on active mode due to all transistors biasing and act as high power consumption class which results into reduction in efficiency. This class of amplifiers consumes power even when there is no output signal. Maximum efficiency is 35% for resistive loads (theoretical value).
- **Class B:** In this class of Amplifiers the transistors can be active only half of the time for increasing the overall efficiency. Although the distortion may be increased [3].
- **Class C:** To increase the efficiency some of the transistors are left unbiased. Here, the conduction angle is smaller than 180 degrees. In this case conduction angle is inversely proportional to the efficiency. Class C amplifiers can be used to amplify the frequency domain modulation.

- **Class D:** In this case at least two transistors must be utilized to support the functioning. In the term of efficiency, it has better performance than the Class C amplifiers. Synchronization between the switches is an issue here to resolve [3, 4].
- **Class E:** This class is mostly used in the RF mobile transmitters, here resonance network is utilized for switching activities. Here efficiency is maximum and timing losses are minimum. Extreme output peak voltages are the problem here [5].
- **Class F:** It can be produced from the combination of square wave voltage of Class D having on switch of Class C. On the basis of shaping of voltage signal into square signal Class F can also obtained from fundamental and third harmonic of the signal from Class B [5].

8.1.3 Heating

Heating is the major problem, when devices working continuously. When a mobile user work continuously, sometimes we found the device get heat up, even in every electronics device we found the same effect due to the same reason that is frequent use of system. Power efficiency here plays a major role to find out the optimal working efficiency. So, one must keep the maximum output power and dissipated power in process of designing of a system. Thermal resistance is a parameter which defines the heating abilities of any electronics system. Thermal resistance of power amplifier depends upon several parameters i.e. layout of the chip, thickness of the chip, thickness of package, die attachment process.

Mathematically, thermal resistance can be defined as

$$\theta t = \frac{T_p - T_c}{P_{dis}} \tag{8.3}$$

Whereas θ_t, T_p, T_c, and P_{dis} represents the thermal resistance, maximum peak junction temperature, case temperature, and power dissipation respectively.

8.1.4 Linearity

Telecommunication standards in general may use the frequency domain modulation, or amplitude domain modulation. When the standards utilize the amplitude domain modulation, then amplifier must be linear condition is required, otherwise the data may be lost. Engineer should keep the tradeoff between the efficiency and linearity with power dissipation while designing the communication system. ACPR (Adjacent channel power ratio) specification nearly defines the linearity.

It can be represented as ACPR(δf).

$$ACPR(\delta f) = \frac{P}{P(\delta f)} \tag{8.4}$$

Whereas *P*, and *P*(*δf*) defines the power transmitted in the channel, and offset frequency of adjacent channel.

8.1.5 Ruggedness

Ruggedness defined as the ability of the system to withstand or work in different electrical failure conditions. For any system, ruggedness can be tested under the output overvoltage, input overdrive and mismatch load conditions. Collector ballasting and emitter ballasting are the two main methods to enhance the ruggedness in Bipolar RF power amplifiers. In the past thicker collector epitaxial region was used to support collector base depletion. But in that method due to large voltage drop, the efficiency was decreases. In present days the emitter ballasting was popularly used for improving the ruggedness in bipolar RF power amplifiers.

8.2 Bipolar PA Design

It is well known that due to the lack of a suitable analytic solution, microwave transistor power amplifiers are actually still designed empirically [1]. Cusack *et al.*'s automatic load contour approach is perhaps one of the most promising large-signal modelling methods to date. This approach to power amplifier optimization is basically a computerized version of the more conventional cut-and-try experimental approach. A characterization of large-signal S-parameters has also made some strides [1].

Monolithic RF power amplifiers have traditionally been dominated by GaAs technology. Recently, attempts have been made to design power amplifiers for the lower frequency spectrum of cell phones using Si technology. The GSM bands at 900 and 1800 MHz are the first targets. Although some architectures are still hybrid, the demand for monolithic power amplifier implementations in traditional bipolar processes is booming. The advantages are self-evident: low cost and simplicity of convergence with other Si-based circuits. In this part, we'll look at a monolithic silicon-based power amplifier. The PCS-CDMA standard, which is widely used in the United States, is the implementation field. The modulation scheme of the CDMA system necessitates a highly linear PA in the handset. Designing a highly efficient monolithic PCS-CDMA PA is incredibly difficult due to the intrinsic high substrate loss and parasitic of Si technologies.

Two common-emitter phases make up the PA (Q0 and Q1). The output-matching network is made up of capacitors C3 to C5, inductors L2, and two transmission lines (CPWs). Capacitors C6 and C7, as well as a transmission wire, make up the input matching network (CPW). Bonding wire field inductors Lgnd1, Lgnd2, and Lgnd3. The inductors L1 and L2 are off-chip inductors. A bonding wire and a PCB trace to form the interstage-matching network link an on-chip capacitor C1 and an off-chip inductor L1. C2 is a built-in bypass capacitor on the chip [6].

Where Rb2 = 15 ohm, the biasing scheme for Q1 produces a nearly constant low-frequency (1 MHz) small-signal impedance at the base of Q1 as well as a nearly linear modulation of Q1's collector's quiescent current. The capacitive high signal RF impedance presented at the base of Q1 at 1.9 GHz is much greater than its counterpart is when Q0 and Q1 are distorted into class AB service. The PA's output power as a function of the RF input signal values. At approximately 600 mW production power, the two-stage PA starts to saturate. Since the average voltage along the bias impedance decreases as the output power rises, this is the case. This, in turn, causes a drop in Q1's base-emitter voltage, forcing it into saturation as a result.

In this section, we'll look at a simplified schematic of a power amplifier (PA) that biases Q1 using an impedance controllable biasing scheme. It's similar to the PA except for the Q1 biasing scheme. Two current-mirror sub circuits make up the biasing scheme: one with transistors Q2, Q4, and Q7, and the other with transistors Q5 and Q6. This biasing scheme is capable of providing independent power of Q1's bias impedance and class of operation if I class and I bias are correctly chosen. I bias controls the output impedance of the bias circuit, while I class controls the output stage's quiescent current. The proposed biasing scheme necessitates optimizing the output stage for efficiency and linearity.

By properly considering the Rd2 and Rd, leads to

$$\mathrm{Vbe(Q1) = Vbe(Q4)}$$

When

$$V(Rb2) = V(Rb3)$$

The current flowing in Q6 must also flow in Q5 and Q4 so Q5 and Q6 form a current mirror. Since the I class controls the current flowing in Q6, it also controls the quiescent current in Q4, which controls the output transistor Q1's quiescent current. In a similar vein, the mechanism of I bias controlling the output impedance of the biasing circuit can be illustrated. By correctly scaling the emitter area ratios between transistor pairs, we can conveniently track the output stage's quiescent current and the bias circuit's output impedance. This allows for the output stage's efficiency to be maximized while maintaining the necessary linearity [6].

Q1's collector quiescent current and the impedance-controllable biasing scheme's low-frequency (1 MHz) small-signal impedance shown at the output stage's base as features of I class and I bias. For the sake of convenience, Rb3 is set to zero (Rb3 = 0). We can see that the control current I class has a similar effect on Q1's quiescent current and the bias impedance in both biasing schemes: the bias impedance is nearly constant, and the quiescent current is regulating almost linearly. Furthermore, by just slightly modify the quiescent current; the control current I bias adds another way to adjust the bias impedance.

The large-signal impedance at 1.9 GHz at the bottom of Q1 is more important. As a function of the input signal frequency, the output power and RF impedance of the impedance controllable biasing scheme. The sizes of Q0 and Q1, as well as the bias parameters, have been set to be similar to those mentioned above for the PA for a logical comparison. There are many advantages of the impedance-controllable biasing system. For examples, the foundation of Q1 has a low impedance. Second, its large-signal impedance is inductive, and its low-frequency small-signal values are identical. By cancelling a portion of Q1's capacitive base impedance, this increases inter-stage matching.

8.3 CMOS Power Amplifier Design

In present scenario, Metal oxide semiconductor plays an important role in comparison to Gallium Arsenide (GaAs) and Gallium Nitride (GaN) technologies due to having more advantages. CMOS innovation empowers activity at a low power supply and a limited creation cost on account of the smaller chip size. CMOS offers the possibility of incorporating radio frequency (RF)/computerized/simple capacities on a solitary chip with ease. Due to low power consumption, low-cost design and being highly integrable, CMOS Power Amplifier plays an important role in modern wireless communication system [7]. The other applications of CMOS PA are in automotive electronics, terrestrial broadcasting, radio broadcasting, baseband processors, sensors, narrowband IoT, GPS, modern shopper gadgets, TV transmissions, telephones, clinical instruments etc.

8.3.1 Performance Parameters

There are number of parameters which are used to measure the performance of CMOS PA. There exists a tradeoff between various performance measuring parameters which makes the designing of PA complex. Amongst all mostly used parameters are linearity, power gain, efficiency, output power and power consumption.

8.3.1.1 Linearity

Linearity basically deals with the resemblance between the input and output. It means for sinusoidal input; desired output is expected to have the sinusoidal behavior [8]. It is an important factor while measuring the performance of RF power amplifier. In the switching RF PA, passive output network with certain impedance is used to meet the linearity characteristic of output waveform obtained from the PA circuit.

8.3.1.2 Gain

The gain of a power amplifier is described as the ratio of output power delivered to the load to the input power fed in the circuit. The high output power delivered by power amplifier makes the power gain high. The power gain is given by the following equation:

$$G = 10\, log_{10}\left({P_{out}}/{P_{in}} \right) \tag{8.5}$$

High gain value represents increasing amplitude. It is represented in dB [8].

8.3.1.3 *Efficiency*

The efficiency or Power Added Efficiency (PAE) is the ratio of AC power delivered by the power amplifier to the DC power applied at the input while considering the input power. It is given by the following equation:

$$PAE = \frac{(P_o - P_{in})}{P_{dc}} \tag{8.6}$$

Large value of output power delivered results in high PAE [9].

8.3.1.4 *Output Power*

Output power is defined as the power delivered by the circuit to the load. It must be as high as possible for a good performing PA. it is described by the following equation:

$$P_O = {V_o^2}/{2\,R_L} \tag{8.7}$$

High output power denotes the more efficient PA [9].

8.3.1.5 *Power Consumption*

The total power consumption in any power amplifier is given by static and dynamic power consumption. Static power consumption is calculated by using the leakage current in the circuit and the dynamic power consumption is given by switching of circuit at high frequency [10]. Power consumption for any PA should be less because less consumption results in less heat in device which makes the device more reliable and in turn increase the efficiency of device.

$$p_{total} = p_{static} + p_{dynamic} \tag{8.8}$$

8.3.2 Drawbacks of CMOS Power Amplifier

There are certain issues with CMOS like gate oxide breakdown, hot carrier degradation, lossy inductive ac grounds, and losses in resonant circuits used to control the harmonics. The oxide breakdown in CMOS happens due to its low breakdown voltage and power amplifiers are mad to work on high voltage. This makes the CMOS a bit unreliable [11]. To overcome these drawbacks, different architectures with increased efficiency have been discussed.

8.3.3 Design of CMOS Power Amplifier

8.3.3.1 Common Cascode PA Design

In 2005, Ding and Harjani designed a common cascode power amplifier, which removes the gate oxide breakdown and hot carrier degradation in CMOS PA [12]. The advantages of architecture are simple circuit, less power consumption, high efficiency and gain.

In common cascode CMOS PA design, biasing of gate terminal of transistor is done by DC voltage. It is designed by integration of class A and class B power amplifier in parallel form to improve the power efficiency and dynamic output range. The common source configuration of PA and two common gate transistors form cascode structure. The CS configuration provides the linear transconductance which in turn provide the high-power gain of 12dB. Biasing of M1 transistor [12] is done by a fixed voltage of 1.2V. For output impedance-matching, coupling capacitors and LC matching network are used on both sides of architecture. Class A amplifier works for low input level by providing high gain and class B PA works for high input level providing high gain. The amplified output by both PA is combined in current domain with certain isolation which makes the overall gain increase. A power efficiency of 44% is achieved by this configuration.

Another configuration in 2005 was designed by Saari *et al.* [12] by using class E PA to improve the overall stability of the network. M3 transistor is used as a switch to minimize the on-resistance of switch and cascode M2 transistor is used to have more output power of 21.3dB with a supply voltage of 3.3 V, the achieved gain is 14.3dB and PAE is 40%. In 2011, Sira *et al.*, designed modulated cascode CMOS class E PA [13] which uses M3 transistor in cascode with M2 transistor and is driven by a power controlling signal at gate. V_{casc} signal value enables M3 to work either in saturation or linear region. A low supply voltage makes to work M3 in saturation while a voltage of upto 1.3 V makes it work in linear region. With a supply voltage of 1.6 V, the achieved PAE is 35%, gain 14.8dB, maximum output power 35dB and 1.2 mm *1.0 mm chip size.

In 2011, Bameri [14] designed another cascode CMOS class E PA which uses the advance power control circuitry. The output power is controlled by the gate voltage of M4 transistor which enables PA to work more efficiently. When the controlled gate voltage decreases, output power also decreases which in turn shift power amplifier into linear mode from the switching mode. On the other hand, when the controlling voltage increases, output power increases which in turn increase the gain as well as efficiency. It gives a 155dB (from -136dBm to +19dBm) power control range. In 2013, Sahu and Deshmukh proposed [15] another design for CMOS PA which is a two-stage design. Both driver and power stage use the MOSFET-only bias which results in reduction of total DC current to 0.0901A. The input and output matching networks are used to reduce the return losses. With a voltage supply of 2.5V, it gives 44.67% PAE and 0.2253W low power consumption.

In 2016, Cai proposed another CMOS PA [16] which uses the two active switches works alternatively. It gives a PAE of 50% while delivering the output power of 15dBm to a 50 Ω load.

The problem with common cascode structure is that it causes more power losses due to parasitic inductance.

8.3.3.2 Self-Bias Cascode PA Design

The low breakdown voltage value of CG transistor configuration put limits on the supply voltage in common cascode structure. In self-bias cascode, the bias voltage is provided by a resistor and a capacitor and higher voltage than V_{DD} is obtained at CG before the oxide breakdown as the architecture allows the swing of RF signal at CG. It reduces the need of extra DC bias which in turn makes the circuit design simple. In architecture, two devices are connecting in series which also reduces the voltage load of each device [16].

In 2009, Murad designed class E self-biased CMOS PA [17] of 0.18 μm technology. The combination of R3 and C3 is providing the bias to the CS gate terminal. The bond wires are used instead of inductors to minimize the power losses which increase the Q factor and in turn increase the PAE. With a supply voltage of 3.3V, the achieved PAE is 44.5% and delivered output power is 23dBm.

In 2010, Hong designed another CMOS PA by using cross coupled capacitor technique [18]. It uses the two stages of amplifiers; first stage uses the common cascode class A PA and the second stage uses the self-biased cacode class AB PA. It implements the cross-coupling structure of capacitors to reduce the by-pass capacitors layout and the large capacitance at gate-drain terminal. The large capacitance is due to the thick gate oxide used in transistor. With a power supply of 3.3V, the achieved PAE is 34.3%, delivered output is 25.2dBm and the power gain is 26.5dB.

The low slew rate of self-biased structure is its major drawback as with rapid changes in input, its output changes very slowly.

8.3.3.3 Differential Cascode PA Design

The differential cascode CMOS PA design has several advantages. It decreases the losses during impedance matching, reduces the common mode noise and substrate coupling, and solves the problem of low breakdown voltage. The architecture uses the cross coupled capacitors to reduce the power dissipation in power amplifier. Capacitors cross coupled network make the positive feedback configuration which is useful to improve PAE but also makes the design of differential cascode PA complex [16].

The differential cascade CMOS PA uses the common source and common gate amplifier configuration. In 2016, Ghorbani designed differential PA by using the cross coupled capacitances neutralization. With a power supply of 1.8V achieved gain is 35.6dB and output power deliver to load is 28.5dBm [16].

8.3.3.4 Power Combining PA Design

When the requirement of output power delivered to the load is high, the single PA can not solve the purpose. To meet the requirement, power combined PA design was invented. Several PA stages are combined together to deliver the high output power to the load. It makes the design more complex which increase the fabrication cost and chip area also [16].

In 2015, Ren proposed another PA, which has one driver stage, one power stage and one parallel combined transformer. Class A and class AB biasing are used by driver and the power stages. The implemented self-biased cascode technique gives ease to the excessive voltage stress to the devices. The PCT gives the compact layout of transformer by combining the three power amplifier stages in parallel [19]. The achieved PAE is 29%, output power delivered is 30.7dBm and gain is 33.2dB.

8.4 Linearization Principles: Predistortion Technique, Phase-Correcting Feedback, Envelope Elimination and Restoration (EER), Cartesian Feedback

The wireless communication network is rapidly growing world-wide due to increasing number of devices connected to the network. The power amplifier plays an important role for transmission of signal as it is used to amplify the modulated signal. Linear power amplifier delivers more output power to the antenna with minimum distortion which makes the linearization of power amplifier an important issue. Linearity refers to the resemblance of the transfer response of input and output to a straight line. Due to non-linear behavior of power amplifier, some undesired frequency signal produced very close to the original signal which is called distortion and very hard to be filtered [20].

When the two or more carriers are used, intermodulation distortion is produced. It cannot be eliminated by using the filter because it is very close to the desired signal frequency. To solve the purpose, various linearization techniques for power amplifiers have been implemented [21]. Figure 8.1 depicts the distortion due to non-linear behavior of power amplifier.

8.4.1 Predistortion Linearization Technique

Pre-distortion linearization is the simplest technique used for the linearization of power amplifier. The idea was to build an analog or digital circuit unit whose input-output power characteristics are just opposite to the characteristics of power amplifier as shown in Figures 8.2 and 8.3. Analog and digital circuit units have their cons and pros. Pre-distortion analog

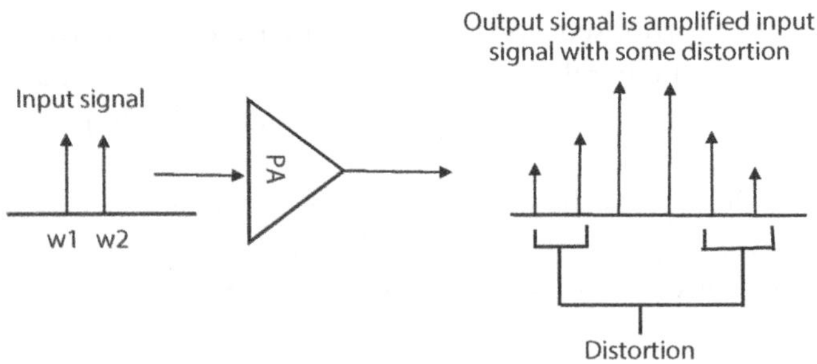

Figure 8.1 Distortion due to non-linear behavior of power amplifier.

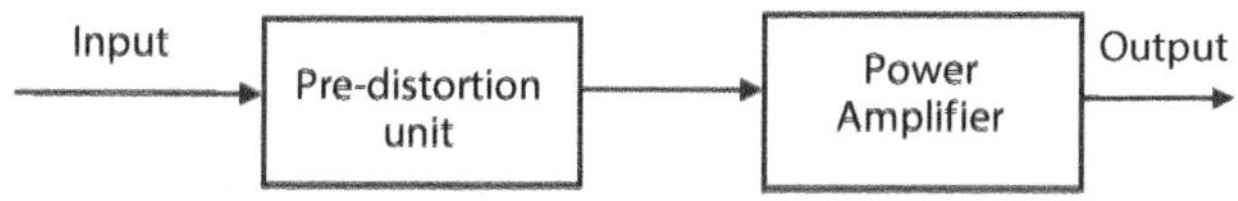

Figure 8.2 Block diagram of pre-distortion linearization technique.

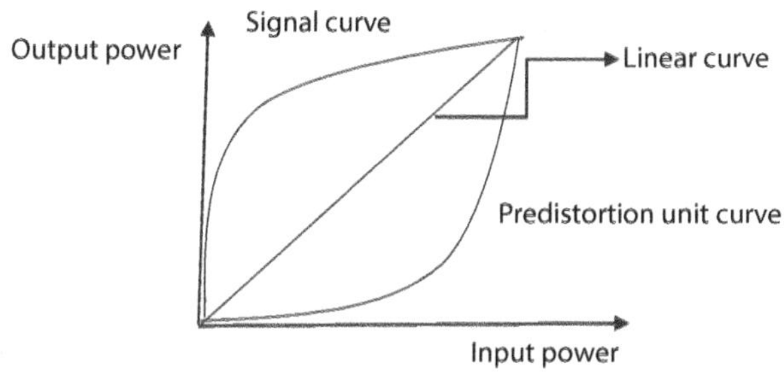

Figure 8.3 Predistortion linearization graph.

unit consumes less power, can be easily implemented and low in cost but it does not provide the full attenuation of distortion and results in the loss of signal power due to insertion of circuit unit (insertion loss). As it consumes less power which makes analog pre-distortion linearization more power efficient [22].

Digital pre-distortion unit overcomes the shortcoming of analog unit. There are various subgroups of digital pre-distortion linearization technique. The most commonly used technique is using the look up table. The look up table consists of premeasured power amplifier output values corresponding to input signal value to attenuate the distortion. Although this technique is simplest but can't be applied everywhere. Also, the ageing of amplifier and temperature cannot be predicted. The second is polynomial based technique where higher degree polynomials are used to determine the behavior of power amplifier.

The adaptive pre-distortion linearization technique makes use of a separate unit known as Transmitter Observing Receiver (TOR), which monitors the output power from the power amplifier and changes the parameter accordingly. The TOR requires more bandwidth which is approximately five times of the input signal bandwidth, thus makes it bandwidth inefficient technique. Also, the energy consumption and cost is high [23].

Another problem with digital pre-distortion technique is the modelling of power amplifier. There are three main categories of power amplifier's model. The memoryless polynomial based model does not have any storage unit, thus deals with amplitude distortion and present signal value. The quasi-memoryless non-linear model also does not have any memory element but it deals with amplitude as well as phase distortion and the present signal value. Non-linear model with memory element has storage unit which can store the past value of power amplifier. So, it deals with present as well as past signal values. The memory unit makes this PA model complex and consumes more power for wide bandwidth.

8.4.2 Phase Correcting Feedback Technique

The phase shift linearization (PSL) is the oldest technique used to get linear power amplifier. It has basically two loops configuration. First loop is used to get the distortion introduced by main power amplifier. Input signal A_{in} is split into two parts by using the power splitter. One part is feed to the main PA to get output signal A_{out}. A_{out} consists of the amplified input signal and some distortion introduced by PA. The other part of A_{in} with gain and phase shift corresponding to that of main PA is introduced to the subtractor. Thus, phase shifted A_{in} is subtracted from the A_{out} to get the distortion only. In the second loop, distortion signal is passed through an auxiliary amplifier which provides amplitude and phase shifted distortion signal. Distorted signal is subtracted from the phase shifted A_{out} signal. Finally, the distortion less signal is obtained at the subtractor output [24]. There are certain shortcomings of linearization technique. While implementing practically, PSL requires the manual setting of amplitude and phase tuning factors. This is because with the time, amplifier raising temperature and ageing effect can degrade the performance if parameters are not

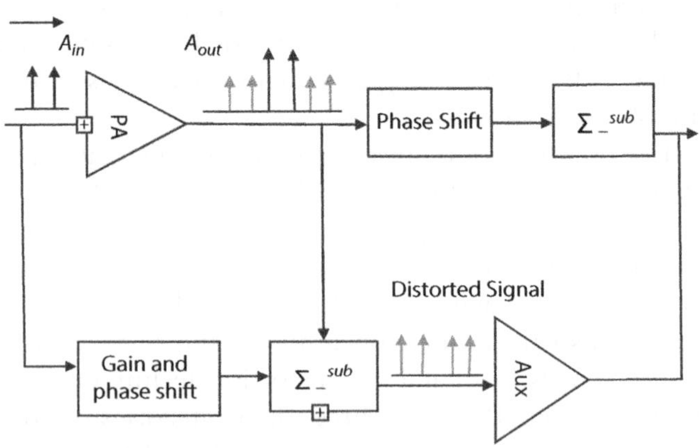

Figure 8.4 Block diagram of phase correcting feedback technique.

set manually. Also, its implementation is complex due to large number of components and it consumes more power. The power efficiency calculated by PSL technique is less. Achieved Power efficiency (PE) is 40% with class C PA and 14% with class A PA. Efficiency decreases due to insertion losses introduced by the gain and phase unit. Due to these shortcomings, PSL is used rarely for linearization [25]. Figure 8.4 demonstrates block diagram of phase correcting feedback technique.

8.4.3 Cartesian Feedback Technique

Cartesian feedback linearization technique (CFLT) is used for linearization of power amplifier. Cartesian domain is considered as compare to polar domain because the decoupling of amplitude and phase of any signal becomes easy in Cartesian domain. In CFLT, quadrature and in-phase property is used where the input signal is divided into two parts with a phase shift of 90. The block diagram of CFLT is shown in the Figure 8.5.

I and Q are the quadrature and in-phase component of baseband signal. It is almost impossible to design a stable, high gain feedback system at high frequency because at high frequency, amplifier introduces lots of oscillations which compromise the stability and gain of system. CFLT considers the two loops, one consists of up conversion frequency translators which are responsible for the modulation of input signal I and Q. The modulated signal is passed to the power amplifier. The second loop consists of the down-conversion frequency translators which is responsible

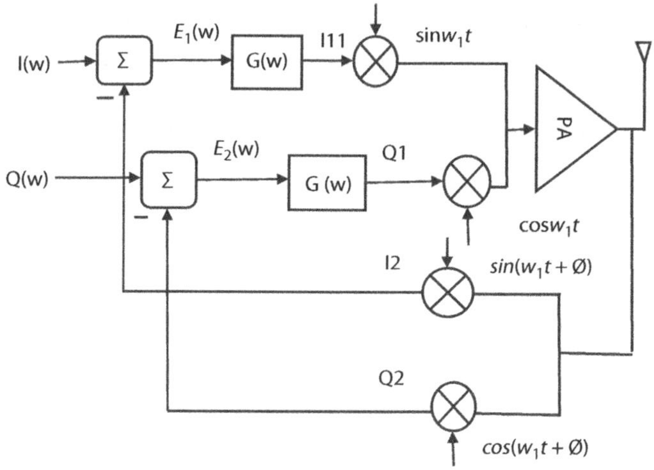

Figure 8.5 Block diagram of Cartesian feedback linearization technique.

for demodulation of output signal produced by the power amplifier. The error signal is calculated by subtraction of the input signal and the signal produced by down-conversion mixers. So, it is the low frequency signal which reaches the baseband signal not the carrier one. In this way, CFLT technique deals with the stability issue. The generated error signal is used to know about the gain and phase shift introduced by the power amplifier. By the means of this information, linearity can be achieved [26].

Advantage of Cartesian feedback linearization technique is that it deals with the temperature and ageing problem of amplifier itself. Due to these factors, does not degrade the performance. But phase alignment is an important issue in CFLT. The whole performance of CFLT is based on the fact that the two loops used for decoupling of I and Q are identical loops. This is possible when phase ø is zero which is an ideal case but practically impossible. In practical, some delay is introduced by power amplifier, due to antenna's reactive load some phase shift is introduced by the RF carrier which makes the phase non zero [27]. So, system stability is affected by the phase misalignment problem.

8.4.4 Envelope Elimination and Restoration Technique

Envelope Elimination and Restoration (EER) technique is used to perform the linearization of power amplifier. It was invented by L. R. Kahn in 1952, thus known as Kahn's technique [28]. The block diagram of conventional EER is shown in Figure 8.6.

In conventional EER technique, the envelope detector is used to get the envelope magnitude only and limiter is used to get the phase component of signal by eliminating the amplitude. Then at the output, by using the varying supply of power amplifier restoration is envelope is carried out [25].

With the growing technology, the conventional structure of EER is modified in digital domain also. Basic idea was digitization of envelope

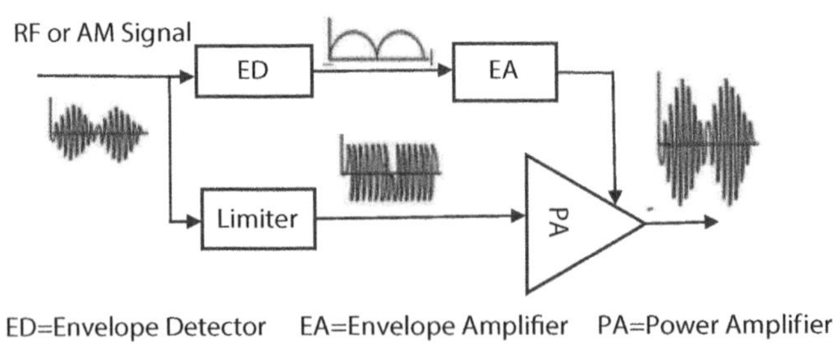

Figure 8.6 Block diagram of conventional EER linearization technique.

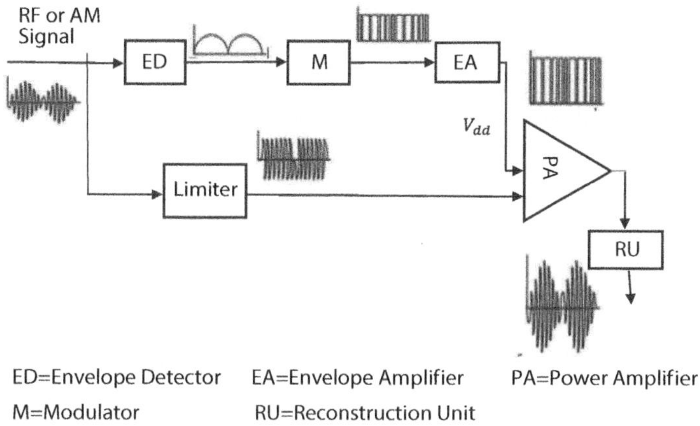

Figure 8.7 Block diagram of digital EER linearization technique.

magnitude (pulse train) and use switching mode power amplifier as a modulator to modulate the constant amplitude phase envelope with the help of generated pulse train. RF input signal which is an amplitude modulated signal is applied to the limiter unit which provides a phase signal of constant amplitude by eliminating the envelope of input signal as shown in Figure 8.7.

The same RF signal is applied to envelope detector which detects the envelope magnitude only. The detected envelope is passed to modulator unit which produce a pulse train of constant magnitude. Further pulse train is amplified by using envelope amplifier which works as supply voltage V_{dd} for PA. The constant amplitude phase signal is modulated by using supply voltage. A bandpass filter is used after power amplifier which restricts the signal to be within desired frequency range only and eliminates all other frequency components. It must be noted that the supply voltage used is binary (either V_{dd} or 0) in nature, thus alleviated all values coming between V_{dd} and 0. All alleviated amplitude and phase values are reconstructed by during the reconstruction of envelope which is last step in EER technique [29].

EER technique is capable of giving linear results by using the non-linear components. Also, it can give 100% efficiency theoretically. In spite of these advantages there are many challenges while implementing EER technique. The path delay for amplitude and phase are different which creates distortion in circuit. Envelope Amplifier (EA) is the main component of EER technique which is responsible for its efficiency but practically it cannot be designed efficiently for wide band.

Acknowledgement

The authors are thankful to Dept. of Electronics and Communication Engineering, National Institute of Technology, Patna for providing resources to write this chapter.

References

1. Krauss, H. *et al.*, *Solid State Radio Engineering*, Wiley, New York, 1980.
2. Raab, F., High efficiency amplification techniques. *IEEE Circuit Syst. (Newslett.)*, 12, 3–11, Dec. 1985.
3. Sechi, F., Linearized class-b transistor amplifier. *IEEE J. Solid-State Circuits*, 11, 4, 264–270, Apr. 1975.
4. Kazimierczuk, M., Class e tuned power amplifier with shunt inductor. *IEEE J. Solid-State Circuits*, 16, 2, 2–7, Feb. 1981.
5. Sowlati, T. *et al.*, Low voltage high efficiency GaAs class e power amplifier for wireless trasnmitters. *IEEE J. Solid-State Circuits*, 30, 10, 1074–1080, Oct. 1995.
6. Kotzebue, K.L., A quasi-linear approach to the design of microwave transistor power amplifiers (short papers). *IEEE Trans. Microw. Theory Tech.*, 24, 12, 975–978, Dec. 1976.
7. Hajimiri, A., California Institute of Technology (Caltech), *188N. Intro. To RF Power Amplifiers*, Aug. 4, 2019. Accessed: April 17, 2021. California Institute of Technology (Caltech), Link-https://www.youtube.com/watch?v=8GLpBE5n5Cw
8. An, K.H., *CMOS RF Power Amplifiers for Mobile Wireless Communications*, Ph.D. Thesis, Georgia Institute of Technology, Atlanta, GA, USA, 2009.
9. Fritzin, J., *Power Amplifier Circuits in CMOS Technologies*, Ph.D. Thesis, Linköping University, Linköping, Sweden, 2009.
10. Sarwar, A., *CMOS power Consumption and CPD Calculation*, Texas Instruments Incorporated, Dallas, TX, USA, 1997.
11. Badal, T., Reaz, M., Bhuiyan, M. *et al.*, CMOS transmitter for 2.4 GHz RF devices: Design architecture of the 2.4-GHz CMOS transmitter for RF devices. *IEEE Microw. Mag.*, 20, 38–61, Dec. 13, 2018.
12. Bhuiyan M.A.S., Badal M.T.I., Reaz M.B.I., Crespo M.L., Cicuttin A., Design architectures of the CMOS power amplifier for 2.4 GHz ISM band applications: An overview. *Electronics*, 8, 5, 477(1-21), 2019. https://doi.org/10.3390/electronics805047
13. Sira, D., Thomsen, P., Larsen, T., A cascode modulated class-E power amplifier for wireless communications. *Microelectronics J.*, 42, 141–147, Jan. 2011.

14. Bameri, H., Hakimi, A., Movahhedi, M., A linear-high range output power control technique for cascode power amplifiers. *Microelectronics J.*, 42, 1025–1031, Sept. 2011.
15. Sahu, S. and Deshmukh, S., Design of high efficiency two stage power amplifier in 0.13 μM RF CMOS technology for 2.4GHZ WLAN application. *Int. J. VLSI Des. Commun. Syst.*, 4, 4, 31–40, Aug. 2013.
16. Cai, W., Huang, L., Wang, S., Class D power amplifier for medical application. *IEIJ*, 4, 2, 9–15, June 2016.
17. Murad, S., Pokharel, R., Kanaya, H. *et al.*, A 2.4-GHz 0.18-μm CMOS Class E single-ended switching power amplifier with a self-biased cascade. *AEU Int. J. Electron. Commun.*, 64, 813–818, Sept. 2010.
18. Hong, J., Imanishi, D., Okada, K. *et al.*, A 2.4 GHz fully integrated CMOS power amplifier using capacitive cross-coupling, in: *Proc. of the 2010 IEEE Int. Conf. Wireless Inf. Technol. Syst*, Honolulu, HI, USA, pp. 1–4, August 28–September 3, 2010.
19. Ren, Z., Zhang, K., Liu, L. *et al.*, On-chip power-combining techniques for watt-level linear power amplifiers in 0.18 μm CMOS. *J. Semicond.*, 36, 9, 095002-1–095002-10, 2015.
20. Borel, A. and Barzdenas, V., A review of basic power amplifier linearization methods. *2020 IEEE Open Conf. Elect. Electron. Inf. Sci.*, Vilnius, Lithuania, June 5, 2020.
21. Katz, A., Linearization: Reducing distortion in power amplifiers. *IEEE Microw. Mag.*, 2, 4, 37–49, Dec. 2001.
22. Ceylan, N., *Linearization of Power Amplifers by Means of Digital Predistortion*, Ph. D. dissertation, University of Erlangen-Nuremberg, 2005.
23. Katz, A., Wood, J., Chokola, D., The evolution of PA linearization: From classic feedforward and feedback through analog and digital predistortion. *IEEE Microw. Mag.*, 17, 2, 32–40, Feb. 2016.
24. Katz, A. and Dorwal, R., Linearizing high power amplifiers. *2012 IEEE Topical Conf. Power Amplifier Wireless and Radio Appl.* Santa Clara, CA, USA, Jan. 15-18, 2012.
25. Parsons, K. and Kenington, P., The efficiency of a feedforward amplifier with delay loss. *IEEE Trans. Veh. Technol.*, 43, 2, 407–412, May 1994.
26. Dawson, J. and Lee, T., Cartesian feedback for RF power amplifier linearization, in: *Proc. of the 2004 American Control Conf.*, Boston, Massachusetts, June 30-July 2, 2004.
27. Dawson, J. and Lee., T., Automatic phase alignment for a fully integrated cartesian feedback power amplifier system. *IEEE J. Solid-State Circuits*, 38, 2269–2279, Dec. 2003.

28. Fedorenko, P., *Phase Distortion in Envelope Elimination and Restoration Radio Frequency Power Amplifier*, Ph.D. Thesis, Georgia Institute of Technology, Atlanta, GA, USA, 2009.
29. Nesimoglu, T., Linearity and efficiency performance of a modified envelope elimination and restoration transmitter architecture. *URSI General Assembly Scientific Symp. Int. Union of Radio Sci.*, 2011.

9

RF Oscillators

Pramila Jakhar[1]* and Amitesh Kumar[2]†

[1]*Department of Electrical & Electronics Engineering, BITS Pilani K K Birla Goa Campus, Goa, India*
[2]*Department of Electrical Engineering, National Institute of Technology, Patna, India*

Abstract
Oscillators are important circuits in Radio Frequency (RF) communication. The main building blocks of an oscillator are amplifier and a frequency-selective network in a positive-feedback. Significant requirement for oscillators used in RF applications should consist of proper amplitude controlled circuitry, low phase error and low power consumption. This chapter presents the important aspects for the design of RC and LC oscillator circuits.

***Keywords*:** Phase noise to carrier ratio, RF oscillators, amplitude control, low power consumption, LC oscillators

9.1 Introduction

The oscillator is one of the important block in RF communication system at both transmission and receiver end. In a transmitter, it generates the signal. The local oscillator is used in conjunction with a mixer in a receiver to transform the received RF signal to an IF signal. A low-frequency analogue or digital signal can modulate the oscillator. The carrier in this example is the oscillator signal, and the information to be sent is the modulating low-frequency signal. High power output, high DC-to-RF efficiency, low noise, strong stability, good frequency tunability, and other characteristics of an oscillator would be desirable in the design of a system.

**Corresponding author*: pramila.jakhar@gmail.com
†*Corresponding author*: dr.amitesh.ee@gmail.com

Sangeeta Singh, Rajeev Kumar Arya, B.C. Sahana and Ajay Kumar Vyas (eds.) RF Circuits For 5G Applications: Designing with mmWave Circuitry, (157–170) © 2023 Scrivener Publishing LLC

A wide number of wireless communication protocols, such as 4G, 5G, and wireless localarea network, are realised and integrated in next-generation mobile devices. Local oscillator frequencies in the GHz range and reference oscillators operating at a few MHz are required by all of these standards. As a result, the demand for oscillators with a broad frequency range is fast increasing.

A negative-resistance solid-state device is at the heart of an oscillator. When subjected to appropriate DC biases, a number of solid-state devices have negative resistance due to device physics. The Gunn device, IMPATT (impact ionisation avalanche transit time) devices, resonant tunnelling diodes (RTDs), and transistors are all examples.

A negative-resistance solid-state device is the heart of an oscillator. Several solid-state devices exhibit negative resistance under specific DC biases due to device physics. The Gunn device, impact ionisation avalanche transit time (IMPATT) devices, resonant tunnelling diodes (RTDs), and transistors are all examples of these.

Figure 9.1 demonstrates the detailed block diagram of RF front end receiver. Figure 9.2 shows the general block diagram of an oscillator.

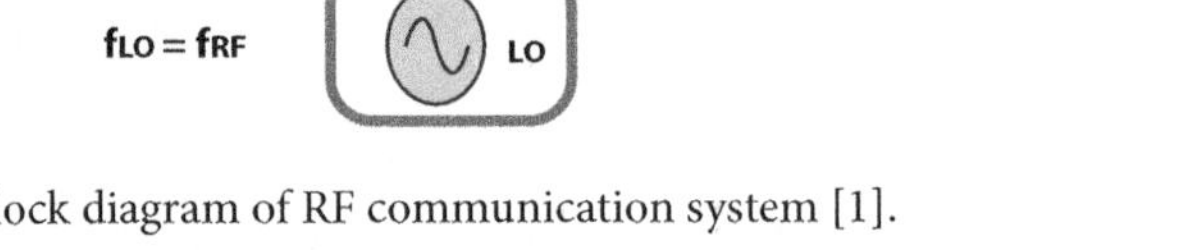

Figure 9.1 Block diagram of RF communication system [1].

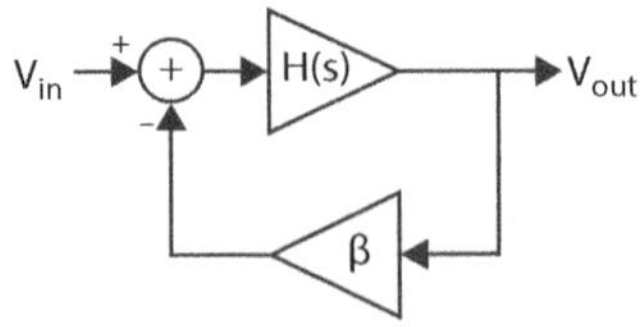

Figure 9.2 Depicts the block diagram of oscillator.

9.2 Specifications

9.2.1 Frequency and Tuning

To think of designing an oscillator, frequency and tuning are very important specifications. Depending upon telecom standards as well as the countries where it would be used, the range varies. Higher frequency waves transmit larger amount of data than lower frequency. There are a variety of micromechanical resonators that can be utilized for oscillators with frequencies ranging from low kHz to GHz. A high-speed switch is also required because the sampling frequency must be significantly lower than the carrier frequency. This simplifies the oscillator design considerably. Some of the disadvantages or considerations to be aware of are noise aliasing, proportionality of the sampling frequency with the carrier frequency, which results in high dynamic-range requirement and high sampling frequency [2]. The frequency of the RF VCO in a wireless transceiver is set by a synthesizer PLL or is unwaveringly regulated by the baseband data across a small fractional range [3].

Tuning specification helps in meeting the limiting conditions for frequency standards. The frequency deviation due to various causes needs to be adjusted in the actual value to help the real-time situations. Some of the configurations of corresponding employed which provides better frequency response with usually wider tuning range. These are considered for making integrated continuous-time filters [4].

9.2.2 Tuning Constant and Linearity

Regardless of whether the oscillator is a Voltage-controlled Oscillator (VCO) or a Current-controlled Oscillator (CCO), any oscillator circuit has a tuning constant. In certain writings, it is also referred to as gain. The tuning range has a direct impact on this constant. As the tuning range expands, so does this constant, and vice versa. The frequency tuning ranges of active inductor-based voltage-controlled oscillators are astonishingly large. A VCO with a wide frequency tuning range was proposed and manufactured using a typical 0.13-m CMOS process by Xu *et al.* [5]. They used a differential active inductor with resistance feedback in their design. This allows for a greater inductance tuning range. A high linearity, high image/LO-rejection in-phase/quadrature (I/Q) direct up-conversion mixer for 5G has been developed by Byeon *et al.* [6]. A high linearity mixer with minimum power amplifier stages would be most appropriate if the goal is to achieve low power consumption with a small chip area. Passive mixers play a crucial role when it comes to achieving great linearity with very low flickering noise [7]. The degradation at in-band frequencies reduces baseband circuit distortion, allowing the auxiliary channel to be turned off. The linearity is greatly improved by turning off noise-cancelling.

9.2.3 Power Dissipation

Consideration of power dissipation is a very important factor in design of oscillator, as it decides the energy requirements and various other related issues involved. In communication circuits as data rate will increases, the complexity of circuit increases which also leads to increased power consumption.

The use of high-speed switch along with very low the sampling frequency than the carrier frequency, is not only helpful in making a simple design oscillator and but also it makes it less power consuming [2]. Continuous Time Linear Equalization (CTLE) was designed in 28nm CMOS technology. Thulasiraman *et al.* have illustrated a tunable and ultralow power consuming active inductor-based for 5G Applications [8].

Shin *et al.* have discussed various modulators where the vector modulators which have active phase shifters with low linearity wideband, however they require high power consumption [8]. Further reflection phase shifters are generally having low linearity and a low power consumption requirement.

9.2.4 Phase to Noise Ratio

Talking about any practical oscillator, amplitude and phase which varies with time, creates disturbance in carrier frequency. Frequency variability in any oscillator leads to phase noises. The change in frequency needs to be adjusted according to the requirement and which is done as per the design used. The frequency domain is where phase noise is classified [9]. A limiter circuit, which may be insignificant, is used to remove the disturbance. This instability is often defined by the phase noise to carrier ratio. Various methodologies, such as jitter and wander constructions, are used to gain a theoretical knowledge of phase noise in oscillators. Staszewski *et al.* have worked on theoretical Oscillator modelling for thorough phase noise characterization of the oscillator. Even time-domain constructs have been suggested to define it [9].

9.2.5 Reciprocal Mixing

As already discussed, phase noise is customarily the primary contributor to oscillator limitations. Considering the characteristics of phase noise, there are various parameters such as jitter, reciprocal mixing & interference.

Reciprocal mixing of the system needs to dominate the performance and phase noise of the measurement equipment need to be further adjusted [10].

Further considering the issues which may arise of reciprocal mixing, we need to consider eliminating RF filtering. Along with gain compression, reciprocal mixing poses a serious challenge in eradicating RF filtering [7]. Desensitization caused by reciprocal mixing of the phase noise of the Local Oscillator by blockers is a serious problem in wideband receivers [7]. However, if the blocker frequency is greater than 400 MHz, the reciprocal mixing problem caused by phase noise has only a minimal impact [11].

9.2.6 Signal to Noise Degradation of FM Signals Spurious Emission

Signal to Noise degradation is another important specification which needs potential consideration. As already discussed about reciprocal mixing and issues which are related to it, any blocker which is otherwise used, need to be down converted in addition with wanted signals. The blockers which when are mixed with LO phase noise, additional noise which is added in form of noise degradation due to reciprocal mixing [7]. Degradation of SNR (Signal to Noise Ratio) has very serious issues and may happen due to clock jitter. This needs a lot many considerations with ADC in its potential design [2].

9.2.7 Harmonics, I/Q Matching, Technology and Chip Area

Harmonics of orthodox mixers are created primarily through non-linearity of the transconductance stage. Any active inductor-based voltage-controlled oscillator (VCO) has an LC-tank topology, which employs cross-coupled transistors as well as the active inductor. It uses fully differential high-speed operational transconductance amplifiers, which remove even harmonics from the output signal. Differential characteristics of the active inductor circuit has a limiting effect on the even harmonics, further considering in-phase/quadrature (I/Q) matching parameter, which can be considered with any I/Q up-conversion mixer which is one of the important components to devise any direct-conversion transmitter. I/Q mismatch as well as LO leakage from the I/Q up-conversion mixer has a very severe effect on degradation in performance (Electron Vector Magnitude performance) of transmitter.

Chip area is further an important consideration. Reducing chip area along with reduction in power consumption requires simpler and competent designs of RF transceivers. These alterations may also help in emerging low power applications like WSN etc. [12]. Using fewer passive components may lead to reduction in gain degradation and thereby reduction in passive loss, which not only reduces power consumption, but also leads to reduction in chip area.

9.3 LC Oscillators

LC oscillators are used in the frequency range of 100 KHz to 100 MHz. The frequency of this type of oscillator circuit is dependent of input control voltage. This frequency is utilized for modulation of signal at transmission side and receiver end.

In this section the basic topologies and important design aspects have been discussed. The basic LC oscillator consist of an inductor L in parallel with a capacitor C which is also called a tank circuit configuration. The electrical equivalent circuit of LC oscillator is represented in Figure 9.3.

The frequency of oscillation is dependent on energy transfer between inductor and capacitor. Initially, the capacitor is charged by the expetrnal energy source and the charged capacitor is connected in parallel with the inductor, thus the charge across the capacitor is then transferred to the inductor which developes a magnetic field. As in parallel L and C circuit there is no other energy storage element so the current through the inductor again charges the capacitor. Thus this process of energy transfer from capacitor to inductor and inductor to capacitor continues and a sinusoidal output voltage with an amplitude and finite frequency is produced.

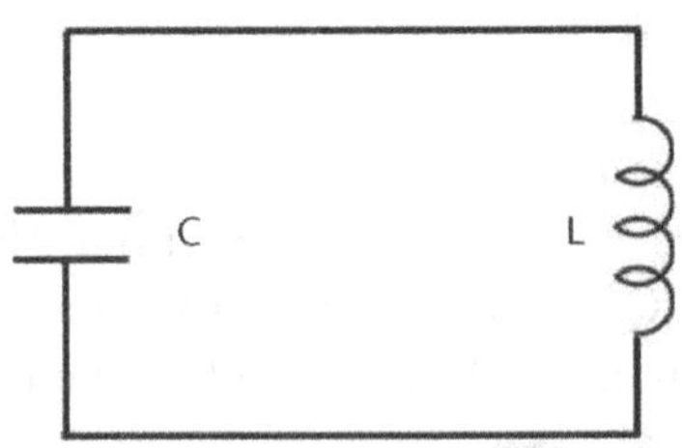

Figure 9.3 Basic LC oscillator circuit.

9.3.1 Frequency, Tuning and Phase Noise Frequency Tuning Phase Noise to Carrier Ratio

One the important design consideration is frequency of oscillation at which LC circuit can produce stable oscillations. For LC oscillators the frequency of oscillation is dependent on inductor and capacitor values which is given by equation 9.1:

$$f = \frac{1}{2\pi\sqrt{LC}} \tag{9.1}$$

The oscillation frequency can be determined by L and C values using equation 9.1. In LC oscillators varactors can be utilized for tuning circuit components. The tuning voltage is used to vary the capacitance of MOS varactors. By varying the DC voltage applied across the MOS varactor can vary the effective capacitance and thus can tune the frequency of oscillation. The varactors can be realized using two NMOS transistors which can function as voltage controlled capacitors.

i. Tuning:

Generally, the tuning in LC oscillator can be obtained by varying L and C. In practice, L is a coil on silicon die with a fixed value and tuning is performed by varying C. The maximum capacitance to minimum capacitance value ratio of varactor can be related with the maximum and minimum value of frequency.

ii. Phase Noise Frequency and Phase Noise to Carrier Ratio:

The first phase model was introduced by Leeson. The formula given by Leeson has several characteristics of real oscillator. Due to the lack of formal proof of formula given by Lesson is considered to be heuristic. Since the formula given by Leeson many reports and phase noise theories have been presented which include non-linear and time-varying aspects in the defining the oscillator model. A linear model can thus be used to calculate the phase noise to carrier ratio of LC oscillators.

The phase noise sidebands of oscillators, in instance, can be split into three areas with differing frequency response slopes. The phase noise with a 6 dB/octave slope, which is generally the greatest region of the oscillator sideband, should be studied first. The 1/f-noise component is the second region. The white noise area is frequently caused by circuits attached to the oscillator, such as buffers.

9.3.2 Topologies

In ideal LC oscillator circuits there are no losses and energy transfer between inductor and capacitor take place which produces sustained oscillation. Though, there are some losses associated with LC oscillator which causes damped oscillations. Thus to compensate for these losses we require to have a negative resistive network of the active circuit components. So in general oscillator circuit it has a frequency selective network consisting of L and C and the amplifier circuit (active component network). The amplifier circuit helps the oscillator circuit to maintain the stable oscillations.

9.3.3 NMOS Only Cross-Coupled Structure

LC oscillator are parallel connection of lossy LC circuit and negative resistive network. This negative resistance compensates the losses in LC circuit and facilitate the stable oscillation by eliminating the resistance associated with the losses in LC tank circuit. The negative resistance can be implemented utilizing the active components and thus the oscillator circuit designed using this is called negative gm-oscillator.

The equivalent electrical resistance between nodes V_x and V_y can be easily determined by using small signal analysis which gives the resistance values as ${}^{-2}\!/_{g_m}$ where g_m is the transconductance of NMOS. Hence, the circuit provides the negative parallel resistance R_p at the resonance condition. To start the oscillation, initially the conductance value is kept negative which satisfy the condition as: $\frac{1}{R_P} - \frac{2}{g_m} < 0$. Thus to start the oscillations R_p should be greater than g_m and in the steady state condition R_p should be equal to g_m. Thus maintaining these conditions is very important for the oscillator circuit design.

1. **PMOS only cross coupled structure:**
PMOS based oscillator circuit design is complementary of NMOS only cross coupled oscillator. In this circuit configuration, the size of PMOS is nearly three times larger than the NMOS to get the same negative resistance due to the higher mobility of the PMOS. The phase noise is less as compared to the NMOS cross coupled structure.

2. **Complementary NMOS PMOS cross coupled structure:**
In this type of topology, both the NMOS and PMOS utilized to from the complementary cross coupled type of structure for oscillator circuit. This

is the most widely utilized as it gives the larger differential voltage swing as compared to the case of single NMOS or PMOS type of transistor. This circuit configuration provides low phase noise. Additionally, the negative resistance is twice for the same power dissipation as the same biasing current flows through the NMOS and PMOS. Hence the total negative resistance of this circuit topology can be determined as parallel combination of NMOS and PMOS negative resistance values which can be given as (equation 9.2):

$$R = -\frac{2}{g_{mn} + g_{mp}} \tag{9.2}$$

Where g_{mn} and g_{mp} are transconductance of NMOS and PMOS circuit configuration respectively. Here, this circuit provides low phase noise and improved signal to noise characteristics.

9.3.4 RC Oscillators

In this section we have discussed some of the practical RC oscillator circuits utilizing active components as op-amp and a frequency selective network consisting of R and C components. These circuits can provide frequency from few Hz to 1 MHz.

Wein-Bridge Oscillator:

It is one of the simplest oscillator circuit. The circuit consist of an operational amplifier in non-inverting configuration with the gain $\mathbf{1+\frac{R_2}{R_1}}$. In the positive feedback the frequency selection network consist of RC network.

The loop gain can be determined as

$$L(s) = \frac{1+{R_2}/{R_1}}{3+RCs+1/sRC} \tag{9.3}$$

From equation (9.3) we can determine the frequency of oscillation $f_0 = \frac{1}{2\pi RC}$.

Oscillations will start at this frequency if the loop gain is at least unity. This can be obtained by selecting $\frac{R_2}{R_1} = 2$.

RC Phase shift Oscillator:

In these circuits the frequency determining conponents are relatively small and cost effective as compared to coils used in LC oscillators. Designers are increasingly turning to this sort of oscillator for the advantages of lower weight and cost while keeping good waveform and stability in the audio range.

An RC phase-shift oscillator's general circuit consists of a negative-gain amplifier (–K) with three RC networks in the feedback. The phase shift introduced by the feedback RC network determines the frequency of oscillation of this circuit. As a result, the total phase shift introduced in the loop at this frequency will be 0° or 360°. At ω_0, the total phase shift introduced by three RC stages will be three times that of one RC network, resulting in a total phase shift of -180°. Operational amplifiers, BJTs [13], or negative resistance devices, JFETs, can be used in the amplifier circuit.

BJT RC Phase-Shift Oscillator:

BJT is a current amplifier. Hence, it is easy to represent the loop gain as a current ratio $A_I(s) = \frac{I_b'}{I_b}$ where I_b' is current injected at transistor's base and I_b is the base current in the loop. Using the small signal analysis the current ratio can be easliy determined and thus sustained oscillation condition can be satisfied. The condition for oscillation be providing the oscillation frewuency as given by equation 9.4:

$$f_0 = \frac{1}{RC\sqrt{6 + 4\frac{R_c}{R}}} \tag{9.4}$$

The JFET RC Phase-Shift Oscillator:

JFET is a high input impedance amplifier circuit. As the JFET has input impedance, so no loading effects will be in the gate circuit. Assuming that the RC network doesn't load the JFET's output and it satisfies the Barkhausen's criterion. According to the sustained oscillation condition, the frequency of oscillation can be given as:

$$f_0 = \frac{1}{2\pi\sqrt{6}RC}$$

9.4 Design Examples

In this section the design examples of LC and RC oscillators are discussed.

9.4.1 830 MHz Monolithic LC Oscillator Circuit Design Measurements

The Silicon on Anything technology is used in the design of an LC oscillator. It can achieve a low phase noise to carrier ratio with a dissipation of less than 1 mW. The common mode rejection ratio of a balanced oscillator circuit might be very high. If the VCO is integrated with other subsystems, like in a trans receiver system, a high common mode rejection ratio is necessary. Because the inductor dominates the balanced LC oscillator circuit design when used with integrated circuits, an increase in chip size is one of the key issues. The on-chip inductors had a value of 29.8 nH in this design, with a maximum quality factor (Q) of 17.3 at 1 GHz. This coil had a total parasitic capacitance of 150 fF.

In addition, seven varactor sections have been connected in parallel, each having a value of 240 fF. For a reverse bias voltage ranging from 0 to 5 V, the maximum to minimum capacitance value was about 1.7. Furthermore, at 1 GHz, the Q value of a PN junction varactor is roughly 26. To achieve maximum tuning range, a typical supply voltage of 5 V was used. The power dissipation of the VCO can be minimized by lowering the value of VCC, but this comes at the cost of a smaller tuning range. The loading of the VCO core owing to the buffer triggered one of the two simultaneous oscillations in this VCO design, resulting in a distorted VCO output signal.

Additionally, the total power dissipation of output buffer was 7.2 mW. When voltage id varied from 5 to 0 V, corresponding measured tuning range is 668 MHz to 830 MHz. Moreover, For the tuning voltage range from 5 to 0 V, CNR (100 kHz) varies between 98.1 and 100.4 dBc/Hz.

9.4.2 A 10 GHz I/Q RC Oscillator with Active Inductors

The use of two integrators in the oscillator circuit allows for a wide tuning range. Two integrator oscillator circuit design with CNR of 106 dBc/Hz at a 2 MHz offset (using 100 mW) and tuning range of 0.9 GHz to 2.2 GHz [15]. Other circuit configurations are more suitable at high frequencies than the RC oscillator. At the penalty of tuning range loss, this surpasses the two-integrator oscillator in terms of phase-noise to carrier ratio and maximum oscillation frequency. The active devices that give the phase noise features of the RC oscillator are used to realize the inductors in this circuit. This can be described as an inductor with a near-unity quality factor. Lumped capacitors are used in the circuit. The amplitude defining requirement for oscillatory activity was made easier by the differential pairs and level current. When the base and collector of transistors are coupled together, inductors can be created, and this two-terminal device can be used as an inductor at high frequencies [16]. Either the base resistance or the current passing through the transistor can be changed to tailor the inductance. This can be used for cascade buffering by providing a load resistance RL in the transconductance collectors. When circuits with low impedance can be linked to the collector of a transconductance, this can produce a no-loading state in the oscillator core. Variable tuning resistance connected to the transconductance base can be used to tune the system.

The variable tuning resistance can be implemented by diode-connected transistor, which was AC-coupled to the base of transistor. Here the current can be controlled by differential pair transistor [14].

9.5 Conclusion

In this chapter, various aspects of RF Oscillators have been considered. Primarily design considerations, in relation to various components and their types have been illustrated. Various specifications including frequency, range, tuning constants, noise factors etc. were discussed which define the design of an Oscillator. Among the Oscillators, LC based Oscillators with their parameters have been discussed which included some important topologies. RC based Oscillators along with appropriate examples were later described. Finally, some comparative examples of LC and RC Oscillators have been discussed to have a clear understanding the various features and their related role on performance of these circuits for 5G considerations.

Acknowledgement

The authors are grateful to the Department of Electrical Engineering at the NIT Patna for providing them with the resources they needed to complete this book chapter. Pramila Jakhar would like to express her gratitude to BITS Pilani Birla Goa Campus for providing her with the resources she needed to complete the book chapter.

References

1. Khatri, R. and Chouksey, S., *Design of 2.4 GHz CMOS LC Oscillator for RF front End Direct Conversion Receiver*, 2019, IEEE, 2019.
2. Bronckers, L.A., Roc, A., Smolders, A.B., Wireless receiver architectures towards 5G: Where are we? *IEEE Circuits Syst. Mag.*, 17, 6–16, 2017.
3. Kral, A., Behbahani, F., Abidi, A.A., RF-CMOS oscillators with switched tuning. *Proc. Cust. Integr. Circuits Conf.*, pp. 555–558, 1998.
4. Tang, X. *et al.*, Design of d-band transformer-based gain-boosting class-AB power amplifiers in silicon technologies. *IEEE Trans. Circuits Syst. I Regul. Pap.*, 67, 1–12, 2020.
5. Xu, J., Saavedra, C.E., Chen, G., An active inductor-based VCO with wide tuning range and high DC-to-RF power efficiency. *IEEE Trans. Circuits Syst. II Express Br.*, 58, 8, 462–466, 2011.
6. Byeon, C.W., Lee, J.H., Lee, D.Y., Kim, M.R., Son, J.H., A high linearity, image/LO-rejection I/Q up-conversion mixer for 5G cellular communications. *2015 Eur. Microw. Week*, pp. 345–348, 2015.
7. Murphy, D. *et al.*, A blocker-tolerant, noise-cancelling receiver suitable for wideband wireless applications. *IEEE J. Solid-State Circuits*, 47, 12, 2943–2963, 2012.
8. Thulasiraman, D. and Gaggatur, J.S., A tunable, power efficient active inductor-based 20 Gb/s CTLE in SerDes for 5G applications. *Microelectronics J.*, 95, 104657, 2020.
9. Staszewski, R.B., Fernando, C., Balsara, P.T., Event-driven simulation and modeling of phase noise of an RF oscillator. *IEEE Trans. Circuits Syst. I Regul. Pap.*, 52, 4, 723–733, 2005.
10. Lien, Y.C., Klumperink, E.A.M., Tenbroek, B., Strange, J., Nauta, B., Enhanced-selectivity high-linearity low-noise mixer-first receiver with complex pole pair due to capacitive positive feedback. *IEEE J. Solid-State Circuits*, 53, 5, 1348–1360, 2018.
11. Jiang, J., Member, S., Kim, J., Karsilayan, A., II, Silva-Martinez, J., A 3-6-GHz highly linear I-channel receiver with baseband bandwidth suitable for 5G wireless and cognitive radio applications. *IEEE Trans. Circuits Syst. I Regul. Pap.*, 66, 8, 3134–3147, 2019.
12. Sun, X., Peng, X., Chen, P.Y., Liu, R., Seo, J.S., Yu, S., Fully parallel RRAM synaptic array for implementing binary neural network with (+1, -1) weights and (+1, 0) neurons. *Proc. Asia South Pacific Des. Autom. Conf. ASP-DAC*, January 2018, pp. 574–579, 2018.
13. Smith, K.C. and Sedra, A.S., *Microelectronic Circuits*, Seventh, Oxford University Press, United States of America, New York, NY, 1395.
14. Vaucher, C.S., Leenaerts, D., van der Tang, J., *Circuit Design for RF Transceivers*, vol. 53, Springer, New York, NY, 2013.
15. Van der Tang, J., Van de Ven, P., Kasperkovitz, D., Van Roermund, A., Analysis and design of an optimally coupled 5-GHz quadrature LC oscillator, in:

Phase-Locking in High-Performance Systems: From Devices to Architectures, vol. 37, pp. 301–305, 2003.
16. Van Der Tang, J., Kasperkovitz, D., Centurelli, F., Van Roermund, A., A 2.7 V, 8 GHz monolithic I/Q RC oscillator with active inductive loads. *Proceedings of the 26th European Solid-State Circuits Conference*, pp. 332–335, 2000.
17. van der Tang, J. and Kasperkovitz, D. A 0.9-2.2 GHz monolithic quadrature mixer oscillator for direct-conversion satellite receivers, *1997 IEEE International Solids-State Circuits Conference*, San Francisco, CA, USA, Digest of Technical Papers, pp. 88–89, 1997.

Part III

RF CIRCUIT APPLICATIONS

10

mmWave Highly-Linear Broadband Power Amplifiers

Shalu C.[1]*, **Shakti Sindhu**[2] **and Amitesh Kumar**[3]†

[1]*National Centre for Flexible Electronics, Indian Institute of Technology, Kanpur, India*
[2]*Oorja Technical Services Pvt. Ltd., Indore, India*
[3]*Nextgen Adaptive System Laboratory (NASL), Department of Electrical Engineering, National Institute of Technology, Patna, Bihar, India*

Abstract
In this chapter, we will discuss the designing of millimeter wave based CMOS PAs (Power Amplifiers) which work in broadband spectrum. Here, the fundamentals of power amplifiers designing, and associated difficulties involved in millimeter wave operations are discussed. A few complexities of AB type power amplifier operations *viz.* efficacy at power back-off, main reasons behind amplitude modulation-phase modulation distortions and techniques to have linear PAs are mentioned. Finally, the last section confers the design and layout of a 29–57 GHz (65% bandwidth) amplitude modulation-phase modulation compensated AB class PA designed for 5G phased arrays. Many methodologies are given for understanding cancellation of wideband distortion and load impedance, though the two-way power combining is allowed to enhance the delivered output power.

***Keywords*:** Power amplifier (PA), transistor, radio frequency (RF), power added efficiency, millimeter wave, switching amplifier, harmonic traps, power divider

10.1 Basics of PAs

10.1.1 Single Transistor Amplifier

A power amplifier is an electronic amplifier designed to increase the magnitude of power of a given input signal. The power of the input signal is

**Corresponding author*: shalu268@gmail.com
†*Corresponding author*: amitesh.ee@nitp.ac.in

Sangeeta Singh, Rajeev Kumar Arya, B.C. Sahana and Ajay Kumar Vyas (eds.) RF Circuits For 5G Applications: Designing with mmWave Circuitry, (173–188) © 2023 Scrivener Publishing LLC

increased to a level high enough to drive loads of output devices like speakers, headphones, RF transmitters etc. [1, 2]. Output current as a result of simulation for 42*12 μm/28 nm common source amplifier ended with short ckt vs I/O DC voltage [3]. It is visible from the graph evidently that I_0 depends on input as well as output voltage in non-linear fashion. Output current affects the output voltage. Impedance at the output node is having complete significant current harmonics. Furthermore, it is expected that the output signal will be larger if the input signal is large. Hence, PA design problem does not lead to insignificant solutions. The PA designers profoundly depend on critical measurement setups like load-pull and wide ranges of simulations. These tools are essential and simultaneously it is always vital to begin with the fundamentals to have clear understanding of important problems as well as margins.

10.1.2 Trade-Offs Among Power Amplifier Design Parameters (P_0, PAE and Linearity)

The PA should be capable enough to deliver adequate power nearly to the 50 antennas, even though confirming enough in-band/out-of-band linearity to fulfill the system level necessities. It is always informative to mention to the simplified schematic to get understanding the basic trade-offs of Power Amplifier design. Let's assume, at operating frequency in resonating input-output matching networks, the delivered power to the load R_L will be

$$P_L = V_L,\text{rms}\ I_L,\text{rms} = V^2{}_L\text{rms}\ /R_L = (n\ \text{Vo,rms})^2\ /R_L. \quad (10.1)$$

The parasitic capacitance of Power Amplifiers is resonated and hence there is center tap for DC biasing. These also provide a galvanic isolation along with impedance transformation and differential to the one ended transformation. The impedance transformation is principally relevant in scaled version of CMOS. The minimum size is changed according to supply voltage in order to confirm the consistent operation and avoid breakdown. The maximum linear output voltage is restricted by and proportional to V_{DD} in a PA topology. Hence, current should increase in order to fulfill the need of output power. So, an impedance transformed n/w made with 1:n transformers(Ideal one) is generally required at the output of Power Amplifier. Hence to fulfill the essential current, it is mandatory to increase the transistor size accordingly. At radio frequency, Power Amplifier requires the DC power to deliver power. The power added efficiency (PAE) measures the efficiency of this conversion and can be given as

$$\mathrm{PAE} = (P_L - P_S)/P_{DC} \tag{10.2}$$

This power is dissipated in transistor as well as in lossy matching networks and hence there is a loss in terms of PAE. In one method, with the help of biasing point, as we limit DC power consumption, we get PAE increased. If VDD = 0V, P_{DC} = 0W then the Power Amplifier gets off results zero power delivered. The PA are classified in several categories e.g. class-A, class-AB, class-B and class-C, depending on the biasing point [1, 2]. Here, some of the points are to be taken care of. Output signal of type-A PA cannot be compressed which results in limitation on maximum input voltage applied, which in turn creates ambiguity on the practical existence of these PAs. In type-C PA, biasing current should be lower than threshold of 0 A. The PA efficiency can be improved by reducing the dissipated power in the transistor. These prevailing techniques have some limitations such as the current should be substantial in order to attain appropriate voltage shaping the harmonic part. Secondly, a higher order tank having high Q-factor, is generally required to achieve a harmonic tuning normally cannot be possible in case of bulk CMOS. Another limitation is regarding harmonics present in amplifiers at their output as those are alternative to the distortion. It sometimes possesses the doubt regarding its outcome on the linearity. The above-mentioned conditions result in trade-off for Power Amplifier design and that is between PAE and linearity. Hence, to save DC power, Power Amplifier must be configured in deep type-AB. Secondly, we presumed that non-linearity of the device increases as we approach threshold voltage. Due to this, it generates additional distortion and harmonics. Hence, distortion plays positive role in case of efficiency and opposite in case of linearity. There are different distortion mechanisms which results in distortion cancellation and hence a higher efficiency for similar or even better linearity [4, 5].

10.1.3 Harmonic Terminations and Switching Amplifiers

Any PA is generally intended to work in large signal region. Hence, the harmonics present at output current are more despite of having single-toned input applied. According to previous section, transconductor dissipates power and it reduces the PAE. So, the load Z_L should be such that it can lower the power loss as well as can shape output voltage by exploiting current harmonics. Herein, there is type-B configured transistor, I_{DC} = 0A having V_{IN} as ideal sinusoidal. If input voltage (V_{IN}) > Vth, transistor operate in saturation and the output current (I_O) = $G_m V_{IN}$ and I_O = 0A when $V_{IN} \leq$ Vth [6].

Here it is assumed that the transistor never goes in the triode region and hence output current depends just on its input counterpart. Evidently, in case of no overlapping between I_O and V_O, there will be zero power delivered in device. In this case, there is 100% theoretical efficiency.

In case when load delivers impedance at harmonics, ideal O/P current and voltage waveforms of the active device.

The equation for the O/P current represented as [6]

$$I_0 = | I_{w0} | + | I_{2w0} | e^{j90^0} + | I_{3w0} | \tag{10.3}$$

Here, I/P voltage is in phase with all odd harmonics. It is 90° out of phase with even harmonics. Third Harmonics onwards can be ignored for ease. The expression for output current can be written as

$$I_0 = -| I_{w0} || Z_{Lw0} | e^{jZ_{Lw0}} - | I_{2w0} | Z_{L2w0} | e^{j(90^0 + Z_{L2w0}} - | I_{3w0} || Z_{L3w0} | e^{jZ_{L3w0}} \tag{10.4}$$

In case of high Q-factor, all harmonics of current are tuned at fundamental freq. In this case, ideal type-B operation is attained. When a significant second harmonic component which is 90° out of phase wrt fundamental component is included to o/p voltage, o/p waveform displays a slighter asymmetric form represented. It is basically a fascinating phenomenon where current's 2nd harmonic component is stronger in type-B PA [1]. This phenomenon is influenced in type-J configuration to increase the efficiency of matching n/w keeping maximum efficiency (78.5%). The load impedance shows a capacitive termination at the second harmonic, which can critically help practically. For having relation 90° in the Equation 10.4, the load impedance is required as inductive at fundamental frequency.

Due to above mentioned reason, the o/p voltage and o/p current are not 180° out of phase. Hence, there is increased amplitude modulation-phase modulation distortion in comparison with type-B configuration. The two circuits with class-F^{-1} and class-F, show better performance theoretically. F^{-1} needs higher impedance at second harmonic and lower impedance at third one while class-F requires vice-versa. Both can achieve a high theoretical efficiency, typically 81.6%. Additionally, a class of amplifiers widely adopted in the sub-GHz range are the switching amplifiers, such as class-D and class-E [7]. These amplifiers attain measured efficiencies approximately 100% provided that good switches are present. In order to refurbish linearity other techniques such as out phasing or envelope tracking may be applied [8, 9]. In case of broadband modulation schemes, both schemes are very difficult to realize. Therefore, at millimeter wave frequencies, these amplifiers are not that much useful.

10.1.4 Challenges at Millimeter-Wave

As we know that gain reduces as transistors approches high speed and, the supply voltage is scaled down with minimum channel length. However, for any class of power amplifiers highest linear output power is proportional to VDD. The blend of these two factors outcomes in the contemporary PAE vs frequency at which transisters are being used. It is clear from the that PAE is decreases with the frequency. Furthermore, a partial supply voltage needs a higher impedance transformation, or wide power combining to attain similar specifications on o/p power. Though, we have certain restrictions on impedance scaling.The reasons for thsese restrictions are highly scalled matching networks.

Extended interconnections of such large devices are simply becoming main sources of power loss at mmWave [9, 10]. Another importent reason is that power combiner's insertion loss increases due to increased combining paths. Hence, delivered o/p power decreases as we increase frequency.

There are many design challenges at the transmitter side *viz.* A wideband PA requires the covering of several channels and amplification of wideband signals in order to confirm the robust performance alongside the PVT variations. Second is the modulated signal with more spectral efficiency to ensure more peak to- average-power-ratio. This is generally a challenge to maintain the linearity vs. efficiency trade-off at a given average P_{OUT}. The last challenge is that the digital pre-distortion is not simply appropriate while integrating several Power Amplifiers array wise for handset applications [11, 12]. Furthermore, the wideband modulation schemes are being adopted at millimeter-Wave, posturing vast challenges on their implementation practically. One of them is digital pre-distortion. Second one is envelope tracking. And another one is outphasing techniques.

10.2 Millimeter Wave-Based AB Class PA

It is evident a PA cis onfigured in class-AB configuration while working at millimeter-Wave frequency. Most of the times, at these high frequencies, a switching amplifier can not work nearby to any switch [17]. Besides, wideband modulation schemes being implemented at millimeter-Wave posses so many limitations on their practical applications as discussed in previous section. Any linear amplifier like class-AB, shows noteworthy advantages at the system level and implementation of Power Amplifiers at millimeter-Wave.

10.2.1 Efficiency at Power Back-Off

A matching n/w at input side comprehends the impedece scaling as 50 and bias voltage is provided with the help of centre tap by this n/w. Matching n/w at the output side is realized using unambiguous 30 resistors along with an inductor L, which resonates output capacitance of the PA. Due to this, centre tap is formed. Tuning of Amplifier is done at 30GHz and transistor is constructed having $(W/L)_{PA}$ = 6*42 μm/28 nm. Traditionally counter-acted CS amplifier is working like transconductance stage. The 1st biasing point is I_D/W = 163.4 μA/micrometer which gives finest ft gm/I_D. Hence, this is a conventional alternative for designing of LNA design. Another biasing point is I_D/W = 28.4 μA/μm. Small inoperative current outcomes many effects. Moreover, the amplitude modulation-phase modulation distortion is 4.8° lower at P1 dB and even worse before P1 dB, where the amplifier is expected to be linear. The notable enhancement in PAE both at P1 dB and at backoff is observed. We noticed a more high harmonic part of o/p voltage. Here, there is no o/p harmonic control. In the next section, there will be the most important phenomena which generates amplitude modulation-phase modulation and circuit linearization solutions.

10.2.2 Sources of AM-PM Distortion

Main reason of amplitude modulation-phase modulation distortion in class-AB power amplifiers are discussed undermany projects [13, 14]. Herein, we have summarized all, alongwith robust emphasis over more pertinent possessions within practical class-AB power amplifiers realized in deep-scaled CMOS. To gain more understanding, a class-AB power amplifier trails an ideal linear driver, $W_{DR} = W_{PA/2}$ has been desirable to mention. It is worthwhile to note that the signal V_{IN} is distorted prior to the amplification because of non-linear I/P capacitance (C_{IN}) of power amlifier even though the driver is linear. If C_{GD} is not the cancelled then C_{GD} is reflected at I/P side because of miller effect. It further gets amplified with $(1 - A_V)$ factor and is equal to $(1 + Gm(Ro//R_L))$.

Hence I/P capacitance does not vary linearly with I/P voltage even though C_{GD} is impeccably linear. Since, the signal can be altered prior to amlification due to non-linear input capacitance, hence the power amplifier is visualized similar to cascading of two non linear elements (i.e. C_{IN}

and Gm). So the total third order intermodulation distortion will be created in the presence of a modulated signal, either by the non-linear C_{IN} and amplified by Gm, or by the second order modulation distortion due to C_{IN} amplified by the non-linear Gm. The non-linear Gm is the main reason for amplitude modulation-phase modulation distortion. O/P current is dependent on the non linearly of I/P and O/P voltage both, hence the involvement is not easy to realize. Lastly, it is worthwhile to mention that in millimeter wave the driver has been required generally [18]. These amplifiers will be significantly benefitted from the non-linear behavior not to waste direct current power, particularly in power back-off. Since, the cascading of non-linear steps is harmful for linearity [19]. Hence, and an well-designed solution to the problem is absent.

10.2.3 Distortion Cancellation Techniques

Main reasons for amplitude modulation-phase modulation distortion are discussed in the previous section. It is promising to plan circuit techniques to relieve them upto some extent. Generally, these techniques are aimed to linearize the devices, De-Qing of I/P RC product and, adding harmonic traps to make O/P of a non-linear power amplifier to perfectly linear one.

10.2.3.1 Input PMOS Varactors

The non-linear deviation of I/P capacitance with NMOS transistor are compensated solely by addition a PMOS transistor or varactor at I/P as projected in [20] and [22]. Evidently, PMOS varactor is efficiently pay off nonlinear I/P capacitance of NMOS CS amplifier.

Furthermore, control voltage could have been employed to reward the PVT differences with formulate imprecision in designing phase. At mmWave, whole I/P capacitance is larger and the corresponding I/P resistance is less because of limited quality factor of varactor. Hence, the amplifier is not as much of subtle to amplitude modulation-phase modulation exchange which is impending by the high quality factor interstage matching network. It would be depicted that it consequences to forfeit the gain. When power amplifier is biased in deep class-AB, application of the method is incomplete. Moreover, amplitude modulation-phase modulation distortion removal competely is not possible in this case. Other effects also present which controls the amplitude modulation-phase modulation distortion of power amplifier if biased deep class-AB. The probable reason of this is to work with durable second harmonic component of O/P voltage and in this case, second harmonic trap will be extra operative.

10.2.3.2 Complementary N-PMOS Amplifier

Second method to eliminate distortion is deep-scaled CMOS which has been appropriate for millimeter wave PAs [20]. It is basically a combination of NMOS and PMOS CS amplifiers. These amlifiers are appropriately sized and biased that consequences in CGS distortion cancellation [16]. The technique is mainly influential in deep-scaled CMOS since the PMOS devices have promoted more from scaling than NMOS. Moreover, it has some faults *viz.* it is not promising to supply scaling and PMOS transistor used as amplifiers are not just a non-linear passive components for capacitive compensation To achieve an ideal operation $V_x \approx V_{dd/2}$ is need. This can be attained in several combinations of Vbp and Vbn. For specific V_{bn}, V_x is fairly sensitive to V_{bp}. Hence, feedback loop is provided to the circuit for appropiate operation.

10.2.3.3 Degeneration Inductance

Degeneration inductance could also been introduced to CS amplifier in the linearization process. Firstly, this metod was offered so that soft compression can be enhanced to conduct millimeter wave power amplifiers [14]. The de-Q the interstage matching network in addition to enhacment in amplitude modulation-phase modulation distortion is perfomed [14, 15]. Hence, power amplifier is not much profound for amplitude modulation-phase modulation distortion emerging from high-quality factor inter-stage matching network [16]. Hence, it is assumed more relaible amplitude modulation-phase modulation distortion. It shows the outstanding routine under the modulated signal.

As long transistors were capable of delivering significant gain at operating frequency, the enhancement in of PAE1 dB with amplitude modulation-phase modulation distortion was worse than 2.5% and 0.6° respectively This benefit has been clearer the deep class-AB. The amplitude modulation-phase modulation distortion improved up to ≈4.2° in comparison to traditional CS amplifier, while attaining a extraordinary PAE1 dB. Moreover, the forfeit in the gain as well sufficient and the amplitude modulation-phase modulation distortion has been also signicant earlier to P1 dB. The power amlifiers which influence the linearization process still require considerable power backoff so that it may attain all linearity necessities.

10.2.3.4 Harmonic Traps

It is evident from earlier conversation of previous sections, power amplifier class-AB biased in high signal circumstances, expreses considerable

AM-PM distortion even after the application of linearization process. The main reason of amplitude modulation-phase modulation distortion is 2nd harmonic component he output [3, 19, 21]. This is ultimately present at power amplifier O/P always. In order to linearize the PA further is to introduce the harmonic traps. Trap is a resonant circuits which demonstrates a low impedance path to ground for second harmonic current.The circuit solutions which realizes the aformentioned situations. Phase can be discarded providing the complete impedance at second harmonic is less. Hence, this technique is mainly robust. Its phase is important if value of $|Z_L|$ is high, for power added efficiency and amplitude modulation-phase modulation distortion [2, 10, 17]. The resolution to this problem can be done by introducing a single-ended series LC filters to understand a short at the second harmonic. Since, lower series capacitance corrosponds to larger quality factor of filter. Hence, Q-factor of filter carry out a back and forth among highest attainable second harmonic suppression as well as B.W. Genearlly harmonic traps are available as common as well as differential mode. Second harmonic trap with I/P varactors outcomes < 2.7° amplitude modulation-phase modulation distortion. It is worthwhile to note that the gain is expanded so the power amplifier is considered over high Q-current which results higher complete small signal gain with less amplitude modulation-phase modulation distortion. The characteristic back and forth among second harmonic suppression with attainable B.W. shows several ambiguities over efficiency in techniques where the large RF bandwidth modulated signal needs to be transmitted.

10.3 Design Example: A Highly Linear Wideband PA in 28 nm CMOS

The mmWave frequency bands allotted around the globe for 5G wireless communication schemes. in this condition, a mmWave massive MIMO/phased array system will be largely benefitted from a single high performance wideband PA.

It is evident from the conversation amplitude modulation and phase modulation distortion compensation techniques in complete bandwidth of process are required to enhance in-band and out-of-band linearity (i.e. EVM and ACPR). It has been shown recently that design of 28GHz CMOS PAs with outstanding efficiency is feasible. However, the BW, average POUT and ACPR under modulated signal considerably limit the attainable link distance and simultaneity with adjacent the channels.

The second harmonic in order to improve the power amplifier linear response, attaining outstanding PAE under modulated signal, along with very inadequate B.W. are shown here. The efficiency of harmonics traps enhances in accordance to quality factor which deals with bandwidth directly. B.W. and PAE are compromised on each other. The effects of second order because of the employment to physical layout are also discussed. Fourth order filters based on the transformer are provided in order to gain impedance scaling and power division/combining with permitted wideband operation and little losses at millimeter wave [18]. Since, P-MOS varactors in Gm I/P along wideband inter-stage matching network have been outstanding in minimization amplitude modulation-phase modulation distortion, so it is accepted in this design. The parameters achieved by this prototype are, Psat = 15.1 dBm ± 1.6 dB and |AMPM| <1° from 29 to 57GHz, with maximum PAE of 24.2%. Moreover, the power amplifier delivers 10.1, 8.9, 5.9 dBm P_{OUT} when amplification is performed at 1.5, 3, 6 Gb/s 64-QAM at 34GHz with EVM/ACPR improved over −25 dB/−30 dBc, prior to application of any distortion.

10.3.1 Transformer-Based Output Combiner and Inter-Stage Power Divider

In order attain appropiate large power level with minimal power supply (generally < 1V), CMOS PAs mostly depend over the power combining methods [17]. A two-port magnetic couple resonator. This could further stretched in a way so that 3-port power combiners can be realized with simple circuit transformation. Magnetic coupling coefficient should enhance and the selfinductance of secondary winding should also reduced in order so that parasitic inductance of the interconnections to the RF pads (LPAD) could be compensated. Similarly, 3-port power divider could be realized by simply interchanging the input and output ports [12]. Here impedance scaling does not required. In addition to it, higher quality factor of load ($R_{IN}C_{IN} > R_O C_O$) imposes $k_{int} < k_O$, to limit bandwidth of power amplifier at provided in-band ripple [11]. The schematic depicts the post-layout gain simulations from I/P of I/P of Power amlifier as well as load. Frequency response of interstage PD displays two maximum as per concept. Hands-on understanding of the transformer properties changes with frequency, upto a point. Futhermore, it demonstrates power of conceptual investigation as given here as well as privous works [13]. Eventhough the power combiners and power dividers look connected with the similiar theory with a simple inversion of the input/output ports. These are the main difference that should be simplified. In the power divider single port is providing power and remaining output ports has been terminated with passive elements. On other hand PCs, two I/P ports has been drived by two currents of equal magnitude and phase. Additionally, k_{par} is the dependent magnetic coupling among the two series transformers. Ideally, k_{par} has the values in the range 0–0.2, then combined impedance increases in complete band. To overcome this unfavorable effect is accepted for the design. In comparison to distributed active transformer (DAT) topology [12], this technique permits additional liberty in layout while decreasing k_{par} appropiately on expenditures of more losses. Assumed topology permits to decrease LPAD, results less insertion loss, though realizing less k_{par} than layout imlementation. It is worth while to note that when the least article scope balances then each technology point levies progressively severe proposal comments. Larger metal density should also contented in spaces of minor area, hence higher insertion loss at mmWave on-chip transformers. The transformer layout used herein for PCs permit to provide similar alignment for PA and driver stages both in order to fulfill this important task. If PC is realized with DAT this technique is not simple. Morover, transmission line based matching networks too available in designing of broad-band power amplifier [9, 15]. These networks have less losses then lumped element module applications. Though, the transformer based filters allow less silicon area ingesting. It provides galvanic isolation and shield alongside ESD proceedings with a balance to unbalance translation and normal DC bias of active circuit. So, lumped element components were adopted herein.

10.3.2 More on the kQ Product

Herein, consequences of kQ product over (1) filter frequency outcome, (2) transformer insertion losses (3) back and forths among BW−3 dB and PAE in millimeter wave PAs which influences the transformer based fourth order matching networks. It is expected kQ product of filter ends is high till now.

However, power amplifier O/P and incomplete applied standards of C_{PAD}, C_O challenge the aforementioned assumption upto an extent which is clear from Equation 10.5. The magnetic coupling constant of O/P transformer (k_O) should be considered large in order to manufacture the essential impedance on the 3 dB bandwidth of power amplifier.

$$f_H = f_L\sqrt{\frac{1+|k_O|}{1-|k_O|}} \tag{10.5}$$

In the design example, f_L is retained constant (25 GHz), and ko is varied, by altering location of 2nd resonant frequency of filter. Now assume a incomplete Q-factor for transformer primary and secondary inductors (L_O, P & L_O, S). Instinctively it is expected that kQ product should be high enough to increase k_O as Q-factor of filter has been reduced with transformer loss. The consequences on load impedance after previously designed transformer depicts $Q_P = Q_S = 10$ at 30GHz. Improved f_L and highly attenuated second resonant peak f_H is observed at the extent of load impedance at first resonant frequency. If the kQ product is not sufficient then desired impedance transformation can not be realized. The insertion losses at PA are the another important aspect for the design of matching network.The insertion loss of lossy transformer are [22]

$$PL_{trans} = 20log\left[\frac{kQ}{-1+\sqrt{1+(kQ)^2}}\right] \tag{10.6}$$

where in this case $Q = \sqrt{Q_P Q_S}$, and Q_P, Q_S, Q-factors of transformer primary and secondary windings. PLtrans in contradiction of magnetic coupling coefficient with dissimilar Q is shown in the schematic 10.20 Lower magnetic coupling coefficient suggests more insertion loss. Though, Q-factor is sufficiently large, deprivation of the transformer routine is incomplete. The PAE for complete PA can be calculated as

$$PAE = \frac{P_{O,lin} - P_{IN,lin}}{P_{dc,DR} + P_{dc,PA}} \tag{10.7}$$

where P_O, $P_{A'lin}$ and $P_{IN'lin}$ are defined inwatt. The power added efficiency (PAE) is 14.6% in the narrowband case ($k_{int} = k_O = 0.8$, $P_{Ldiv} = 1.8$dB). PAE reduces by 0.04% if broadband inter-stage PD has been emlyoed ($k_{int} = 0.4$, $P_{Ldiv} = 2.15$ dB). It is clear that high gm and ft available in 28 nm CMOS profited by tradeoff between bandwidth and linearity for a mmWave PA is generally relaxed. Considering enhancements in amlitude modulation-phase modulation and memory possessions over a fine band alternate has been conversed and it has been established that wideband design is desirable.

10.4 Conclusion

The basics of PAs for millimeter wave applications have been conversed within this chapter. Firstly, the most important back and forths and experiments have been demonstrated in section 10.1. The section 10.2 deals with some facts of class-AB amplifier behavior at millimeter wave frequency. In this section many reasons for amlitude modulation-phase modulation distortion have been underlined and general contemporary explanations are proposed. Lastly, section 10.3 represents wideband amlitude modulation and phase modulation compensated class-AB PAs that is appropiate in integrated 5G phase arrays. To realize these methods broadband impedance transformation, power division/combining and phase distortion linearization have been elaborated in detail. Practical layout constrains forced by deep-scaled technologies causing second order effects are acknowledged and the modest design methods are suggested. In 28 nm bulk CMOS power amplifier working at 0.9V allows to measure Psat =15.1 dBm±1.6 dB and |AM-PM| <1° from 29 to 57GHz, with a peak PAE of 24.2%. At last the consequences of kQ product over (1) filter frequency response, (2) transformer insertion loss, (3) back and forths between 3-dB bandwidth and PAE in millimeter wave PAs has been discussed.

Acknowledgments

Author Shalu C. is thankful to NCflexE, IIT Kanpur for providing encouragement and necessary resources to write this chapter. Shakti Sindhu is thankful to Mr. Saurabh Pishwe, M.D. Oorja Technical Sevices Pvt. Ltd for encoragement and support. Amitesh Kumar is thankful to Dept. of Electrical Engineering, National Institute of Technology, Patna for providing resources to write this book chapter. Amitesh Kumar is thankful to the Council of Scientific & Industrial Research to provide research fellowship to carry out research during his stay at the Indian Institute of Technology, Indore.

References

1. Cripps, S., *RF Power Amplifiers for Wireless Communications*, Artech House, 2006.
2. Vigilante, M. and Reynaert, P., *5G and E-Band Communication Circuits in Deep-Scaled CMOS*, Springer International Publishing AG, part of Springer Nature, 2018.
3. Cripps, S.C., *Advanced Techniques in RF Power Amplifier Design*, Artech House, 2002.
4. Park, B., Jin, S., Jeong, D., Kim, J., Cho, Y., Moon, K., Kim, B., Highly linear mmWave CMOS power amplifier. *IEEE Trans. Microw. Theory Tech.*, 64, 12, 4535–4544, 2016.
5. Sansen, W.M.C., *Analog Design Essentials*, vol. 859, Springer Science & Business Media, Berlin, 2007.
6. Reynaert, P., Steyaert, W., Vigilante, M., RF CMOS, in: *Nanoelectronics: Materials, Devices, Applications*, vol. 2, 2017.
7. He, Y., Li, L., Reynaert, P., 60 GHz power amplifier with distributed active transformer and local feedback, in: *2010 Proceedings of the ESSCIRC (ESSCIRC)*, pp. 4314–317, 2010.
8. Zhang, H. and Sanchez-Sinencio, E., Linearization techniques for CMOS low noise amplifiers: A tutorial. *IEEE Trans. Circuits Syst. I Regul. Pap*, 58, 1, 22–36, 2011.
9. Shahmohammadi, M., Babaie, M., Staszewski, R.B., A 1/f noise upconversion reduction technique for voltage-biased RF CMOSoscillators. *IEEE J. Solid-State Circuits*, 51, 11, 2610–2624, 2016.
10. McCune, E., Fundamentals of switching RF power amplifiers. *IEEE Microw. Wirel. Compon. Lett.*, 25, 12, 838–840, 2015.
11. Kulkarni, S. and Reynaert, P., A 60-GHz power amplifier with AM-PM distortion cancellation in 40-nm CMOS. *IEEE Trans. Microw. Theory Tech.*, 64, 7, 2284–2291, 2016.
12. Mortazavi, S.Y. and Koh, K.J., Integrated inverse class-F silicon power amplifiers for high power efficiency at microwave and mmWave. *IEEE J. Solid-State Circuits*, 51, 10, 2420–2434, 2016.
13. Golara, S., Moloudi, S., Abidi, A.A., Processes of AM-PM distortion in large-signal single-FET amplifiers. *IEEE Trans. Circuits Syst. I Regul. Pap.*, 64, 2, 245–260, 2017.
14. McCune, E., A technical foundation for RF CMOS power amplifiers: Part 5: making a switch mode power amplifier. *IEEE Solid-State Circuits Mag.*, 8, 3, 57–62, 2016.
15. Zhao, D. and Reynaert, P., A 60-GHz dual-mode class AB power amplifier in 40-nm CMOS. *IEEE J. Solid-State Circuits*, 48, 10, 2323–2337, 2013.
16. Chappidi, C.R. and Sengupta, K., Frequency reconfigurable mmWave power amplifier with active impedance synthesis in an asymmetrical non-isolated

combiner: Analysis and design. *IEEE J. Solid-State Circuits*, PP, 99, 1–12, 2017.

17. Zhao, D. and Reynaert, P., An E-band power amplifier with broadband parallel-series power combiner in 40-nm CMOS. *IEEE Trans. Microw. Theory Tech.*, 63, 2, 683–690, 2015.
18. Shakib, S., Park, H.C., Dunworth, J., Aparin, V., Entesari, K., A highly efficient and linear power amplifier for 28-GHz 5G phased array radios in 28-nm CMOS. *IEEE J. Solid-State Circuits*, 51, 12, 3020–3036, 2016.
19. Razavi, B., *RF Microelectronics*, 2nd, Prentice Hall, New Jersey, 2011.
20. Vigilante, M. and Reynaert, P., A wideband class-AB power amplifier with 29–57-GHz AM-PM compensation in 0.9-V 28-nm bulk CMOS. *IEEE J. Solid-State Circuits*, 53, 5, 1–14, 2018.
21. Wang, C., Vaidyanathan, M., Larson, L.E., A capacitance-compensation technique for improved linearity in CMOS class-AB power amplifiers. *IEEE J. Solid-State Circuits*, 39, 11, 1927–1937, 2004.
22. Ye, W., Ma, K., Yeo, K.S., 2.5 A 2-to-6GHz class-AB power amplifier with 28.4% PAE in 65nm CMOS supporting 256QAM, in: *2015 IEEE International Solid-State Circuits Conference-(ISSCC) Digest of Technical Papers*, San Francisco, CA, pp. 1–3, 2015.

11

FinFET Process Technology for RF and Millimeter Wave Applications

A. Theja[1*], Vikas A.[1], Meena Panchore[1] and Kanchan Cecil[2*]

[1]Department of Electronics and Communication Engineering, National Institute of Technology Patna, Bihar, India
[2]Department of Electronics and Telecommunication Engineering, Jabalpur Engineering College, Jabalpur, Madhya Pradesh, India

Abstract
FinFETs are considered as a potential candidate for modern complementary-metal-oxide-semiconductor (CMOS) technology due to reduced short channel effects (SCEs) and better gate controllability over channel. However, with the downscaling of device dimensions, the RF performance at higher current levels is degraded. Therefore, in this chapter, we have presented the scaling effects on RF performance of Fin devices including the parasitic and noise components. This chapter also focused the impact of self-heating and temporal process variability on electrical performance of Fin devices.

Keywords: Fin, scaling, self-heating effect, RF performance, millimeter-wave (mmWave)

11.1 Evaluation of FinFET Technology

In planar CMOS transistors, the speed of the device has been increased due to scaling of device dimension. However, the device scaling at sub-micrometer deteriorates the electrical characteristic and affects the analog and RF performance at circuit level. Hence, FinFET is considered as a potential candidate to overcome the limitations of planar devices at sub-nanometer regime. This section briefly covers the limitations of planar bulk CMOS

**Corresponding author*: at.phd20.ec@nitp.ac.in

Sangeeta Singh, Rajeev Kumar Arya, B.C. Sahana and Ajay Kumar Vyas (eds.) RF Circuits For 5G Applications: Designing with mmWave Circuitry, (189–222) © 2023 Scrivener Publishing LLC

transistor over FinFET technology and explains the optimization techniques at millimeter wave applications. Further, the fabrication steps are discussed to build FinFET structure at sub-micrometer node.

Figure 11.1 shows the cross-sectional view of FinFET device. Here, the FinFET gate covers a thin cut of pure silicon, referred as a "fin", and current flows from one end of the fin to the other [1]. The gate controls over the channel, thus avoiding leakage currents and reduces Short Channel Effects (SCEs). The reduced SCE leads to higher intrinsic gain and lower leakage current in the OFF-state condition. The current flows on three facets [2] of the fin, therefore, the equivalent channel width(W) is equal to the sum of width of fin (W_F) and twice of its height (H_F) i.e. $W = W_F + 2H_F$. Typically, width and height are 6 nm and 50 nm respectively. Since, the height of the fin, H_F is not under control of circuit designer, it appears that the width of fin, W_F can be selected such that $W_F + 2H_F$ yields the desired transistor width. However, the width of the fin affects the device characteristics such as drain/source series resistance and channel length modulation. Due to this reason, the width of the fin is also fixed leading discrete values for the width of the transistor. For example, if $W = W_F + 2H_F = 120nm$, then broader transistors can be obtained by increasing the number of fins.

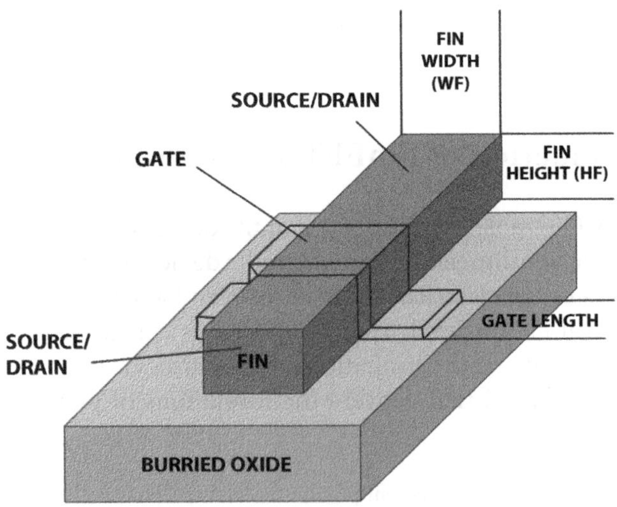

Figure 11.1 Cross sectional view of FinFET.

11.1.1 Steps of Fabrication and Process Flow of FinFET Technology

To understand the device performance of FinFET due to process variations, this section presents the simplified process flow of a FinFET technology as depicted in Figures 11.2 and 11.3. The important steps involved in the fabrication are as follows:

The process starts with the formation of fin and it is done using optical lithography followed by plasma etching. The choice of thickness for the spacer is crucial, as it determines the width of the final fin. After etching process, the sidewalls becomes rough, therefore the process of oxidation and H_2 annealing are performed for the smoothness of the sidewall surfaces. After that, the gate dielectric is developed and then the metal gate is grown. It is preferred to select the V_{th} of the device by using a material (which can also be used to form a gate) that has the appropriate work function instead of doping the channel like in planar MOSFETs. In FinFETs, the channel is usually lightly doped or undoped to suppress SCE, hence,

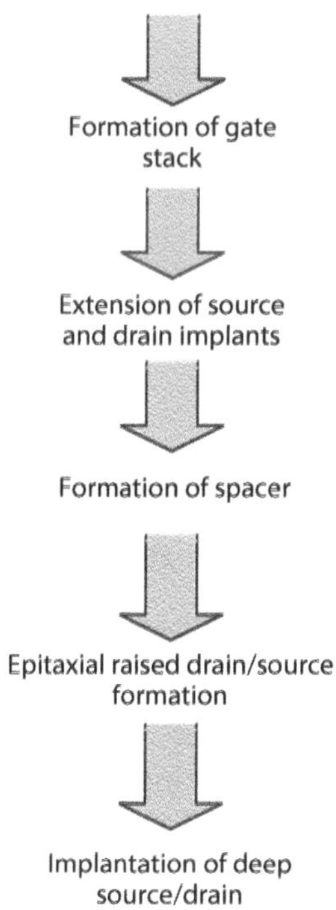

Figure 11.2 Steps involved in fabrication of FinFET technology.

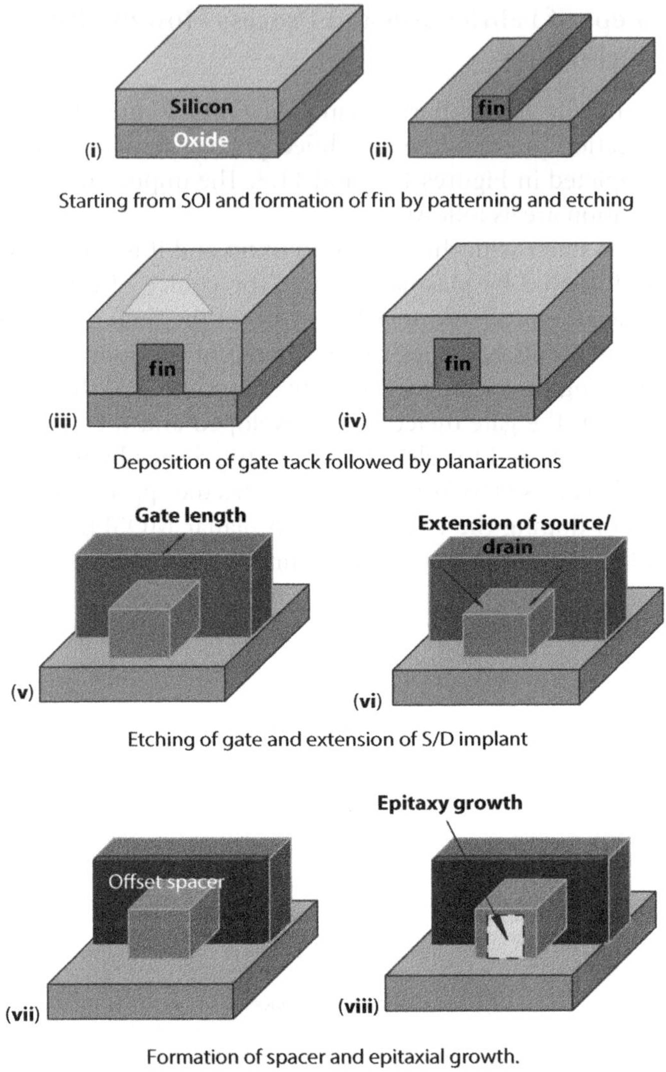

Figure 11.3 Process flow of FinFET technology [4].

the V_{th} is tuned with the help of gate work function. Next step is the gate stacking over the fin, while depositing the gate material. After this step, planarization of the Gate stack is done, so that, the gate etching can be performed. In this etching process, the gate must have a high selectivity to prevent the damage of fin. After this, S/D extensions are developed by using an implantation. Later, source/drain offset spacers are established along the sidewalls of the fin as well as gate.

11.1.2 Digital Performance

In this section, the digital performance metrics like intrinsic transistor performance (ITP), threshold voltage (V_{TH}) and subthreshold slope (SS) at different High-K gate dielectric [3] of planner MOSFET and FinFET are discussed.

Figure 11.4 shows the ITP of both devices in terms of I_{ON}/I_{OFF} ratio at different High-K dielectric materials. Here, I_{ON} defined as the saturation current, and I_{OFF} represented as the total leakage current, which is the combination of junction leakage, gate leakage and subthreshold leakage currents. By observing the ITP curve, we can say that the planar bulk transistors with SiON dielectric exhibit large OFF-state current as compared to planar MOSFETs with High-K dielectric. The high OFF-state current in planar bulk devices is because of the gate leakage through the SiON dielectric. This also reduces the maximum drive current due to mobility degradation. On the contrary, in FinFETs, the leakage current reduces due to the presence of buried oxide (BOX) layer and enhances drive current due to strong control of gate over channel.

Figure 11.5 shows the comparison of threshold voltage (V_{TH}) of both the devices in saturation region. As observed from the figure, FinFETs exhibit

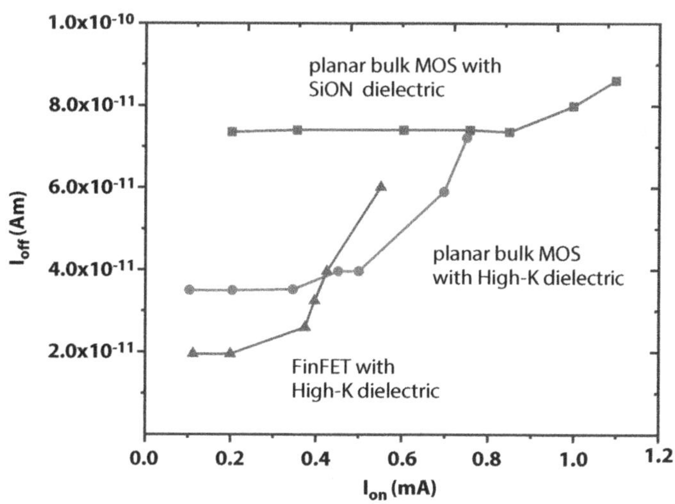

Figure 11.4 ITP performance of FinFETs and planar MOSFETs.

lowered values of V_{TH} (~0.2–0.4 V) compared to bulk planar MOSFETs, which is significant for lower technology nodes. The reduction in V_{TH} of FinFETs is due to the existence of BOX layer and the undoped nature of the fins, which further reduces the voltage drop across the semiconductor,

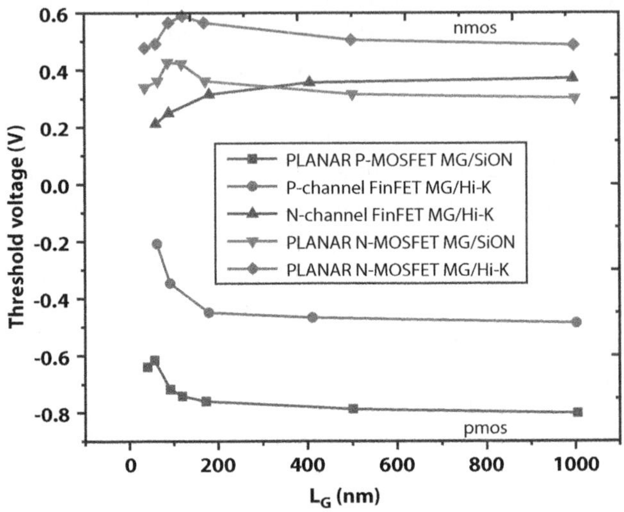

Figure 11.5 Threshold voltage of FinFET and planar MOSFET at different High-K gate dielectric.

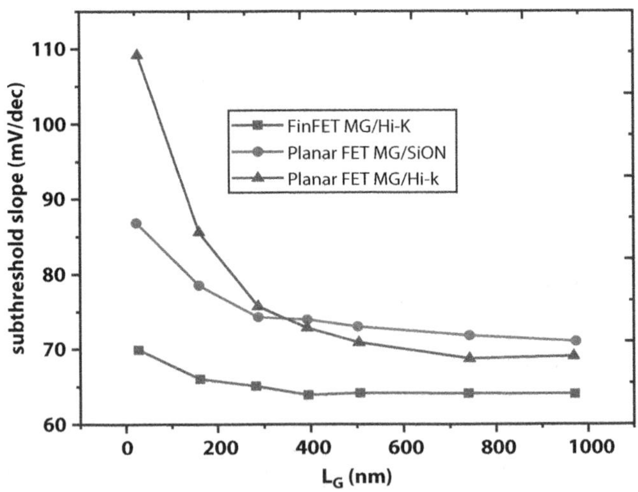

Figure 11.6 Subthreshold slope of FinFET and planar MOSFET at different High-K gate dielectrics.

hence the V_{TH}. Further, the FinFET shows excellent performance in terms of SS which is less than 70 mV/decade as compared to the planar MOSFETs as shown in Figure 11.6.

11.1.3 Analog/RF Performance

In this section, the Analog/RF performance figure of merits (FOMs) like transconductance (G_m), voltage gain (A_V), Output conductance (G_{DS}) and cut-off frequency (f_T) of FinFET as well as planar MOSFET are discussed at different High-K gate dielectrics. Figure 11.7 shows the G_m variations with respect to the physical gate length of both the devices. It has been observed that the transconductance of FinFET and planar MOSFET are similar till $V_{GS} - V_T = 0.3$ V.

After $V_{GS} - V_T = 0.3$ V, the transconductance of the FinFET decreases as compared to that of the planar MOSFET due to the high series resistance of FinFET which reduces the drive current and hence transconductance.

The comparison of output conductance (G_{DS}) of FinFETs as well as of Planar MOSFET is shown in Figure 11.8. It is observed that the FinFETs exhibit reduced output conductance than planar MOSFETs. This is because the FinFET is completely depleted due to the limited fin width and the control of the gate over the channel. Hence, increasing the voltage at the drain, more than the pinch-off voltage, does not deplete the body further. As a result, the channel length modulation, which occurs due to the reduction in the channel length on account of the expansion of the depletion region

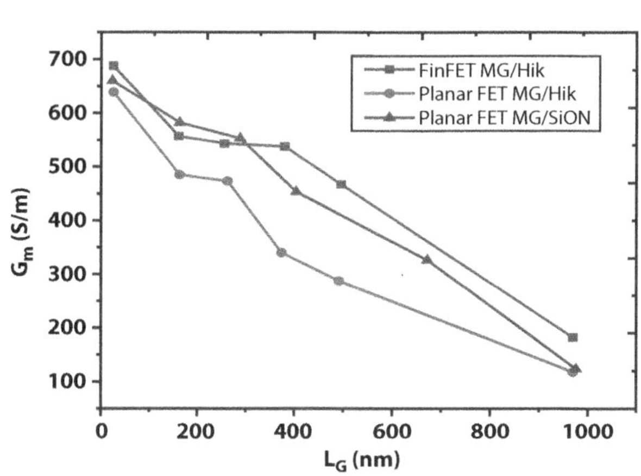

Figure 11.7 Transconductance of FinFET and planar MOSFET at different High-K gate dielectrics.

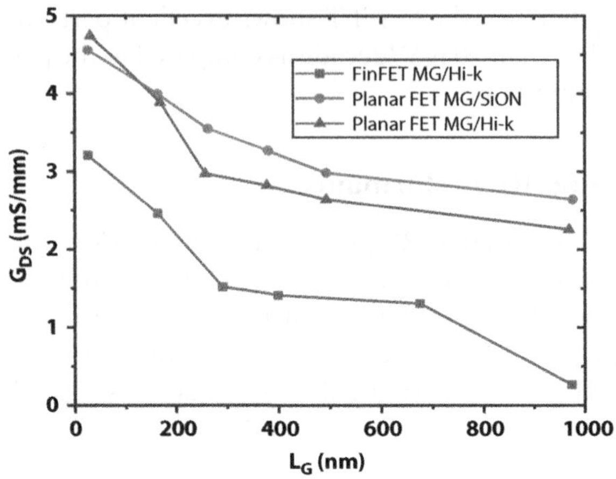

Figure 11.8 Output conductance of FinFET and planar MOSFET at different High-K gate dielectrics.

at the drain, is low. Therefore, the change in drain current with respect to drain voltage, which represents the output conductance is also low.

Figure 11.9 shows the voltage gain (A_V) of FinFET and the planar MOSFET at different High-K gate dielectrics, which is given by the ratio G_m/G_{DS}. FinFET shows higher voltage gain due to high transconductance and low output conductance as compared to the planar bulk MOSFETs.

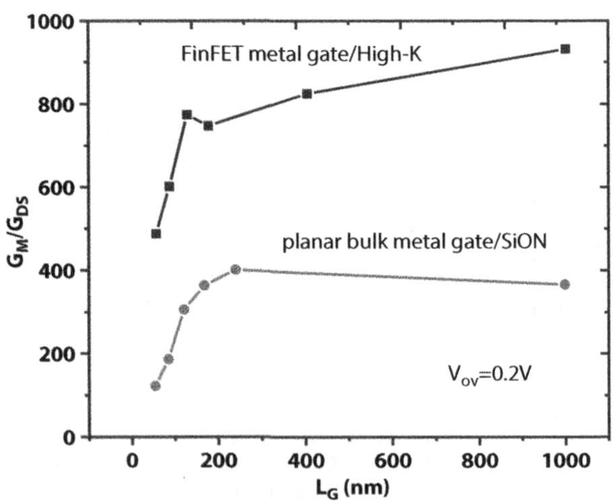

Figure 11.9 Voltage gain of FinFET and planar MOSFET at different High-K gate dielectrics.

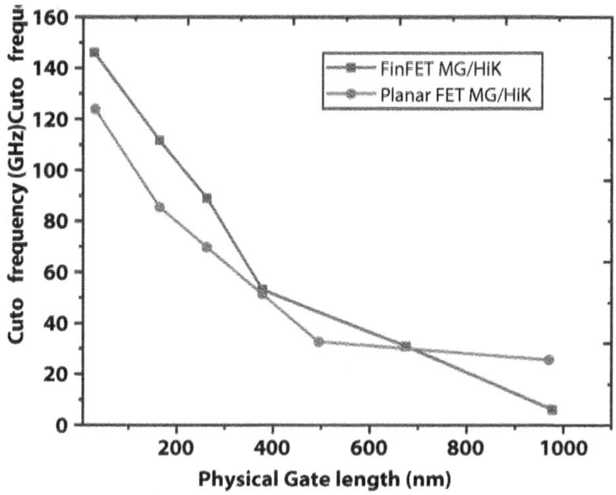

Figure 11.10 Cut-off frequency of FinFET and planar MOSFET at different High-K gate dielectrics.

This is a significant advantage of FinFET over the planar MOSFET for RF circuit applications.

Another parameter that determines the high-frequency performance is the cutoff frequency f_T (= $G_m/2\pi C_{GS}$), which is shown in Figure 11.10. FinFETs and planar MOSFETs exhibits comparable cut-off frequencies because both the devices are having similar values of capacitance (C_{GS}) and transconductance. For analog applications, the other important parameters to be considered are linearity and noise. It has been observed in FinFETs that there is no severe deterioration either in linearity or noise with respect to planar MOSFETs. Therefore, it can be concluded that for analog/RF design, FinFETs exhibit similar value of cut-off frequency and large voltage gain as compared to planar MOSFETs with same linearity and noise levels.

11.2 Distinct Properties of FinFET

In recent time, the requirement of Nanoscale devices with enhanced performance is essential than the conventional CMOS devices. It has been observed that the new emerging devices with add-on features are more suitable for RF/mmWave applications. Hence, this section briefly describes some unique properties of FinFET technology which makes it suitable for RF/mmWave applications.

11.2.1 Performance with Transistor Scaling

For the past five decades, Moore's law has been sustained through the frequent reduction of transistor channel length which in turn increased transistor density and performance. In the case of planar bulk MOSFETs, the performance improvements have reached saturation level at lower technology nodes, forcing the semiconductor industry to shift from planar to tri-gate structures [5]. The shorter channel length of the transistor leads to the higher transconductance (g_m) scaled by the square of channel length (L^2) and the lower gate capacitance, thus higher unity gain frequency (f_T). However, as the channel length is decreasing day by day, the transistor performance improvement trend has been significantly disturbed. The two major causes for this are velocity saturation and drain induced barrier lowering (DIBL) effect [6].

When gate length is reduced, i.e. the device scales to an utmost level, the source and drain start to interact more through the region below the channel despite of gate potential, and the gate control of the device is reduced. This effect is called short channel effect degradation, and the parameter used to measure this is DIBL. In an ideal device, the potential barrier in the channel is supposed to be lowered only through the gate, which is valid for long channel devices. But this assumption is no longer applicable in case of short channel devices. In FinFETs, the tangential electric field produced at the drain is more strongly screened from the source end of the channel due to closeness of the channel to the second gate, results in suppressed SCEs in particular, better subthreshold swing (SS) and reduced DIBL.

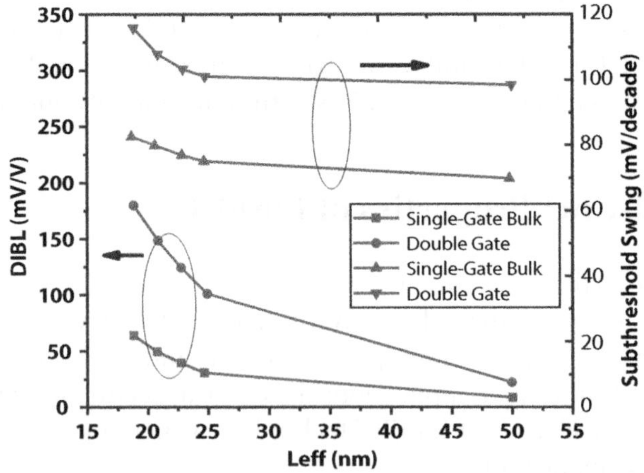

Figure 11.11 DIBL and subthreshold swing of DG and planar MOSFETs.

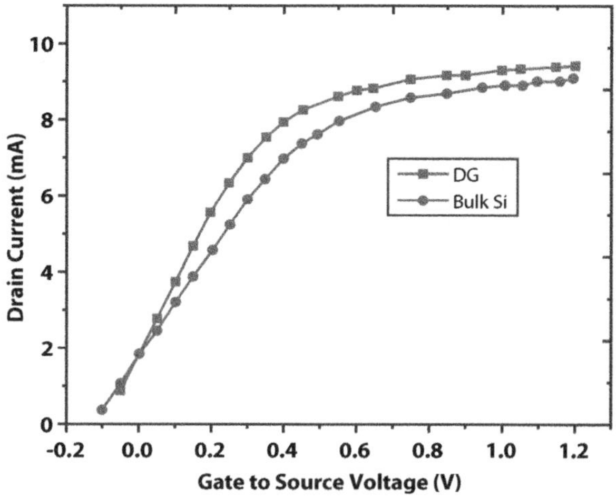

Figure 11.12 Transfer characteristics of DG and planar MOSFETs.

Figure 11.11 shows the Subthreshold swing and DIBL of planar MOSFETs and Double Gate (DG) devices with respect to the effective gate length. It can be observed that both the subthreshold swing and DIBL of the DG devices are significantly enhanced compared to planar MOSFETs.

In planar bulk MOSFETs, DIBL can be reduced by increasing the doping concentration of the body. This in turn also improves the subthreshold swing, thereby desiring higher V_{th} to maintain the low subthreshold current. On the other hand, reducing the doping concentration of body could enhance the subthreshold swing but degrade DIBL. Thus, a compromise is needed for the design of planar MOSFETs.

The transfer characteristics (I_{DS}–V_{GS} curves) of DG and planar MOSFETs are shown in Figure 11.12. From the characteristics, it can be observed that DG-FET has lower threshold voltage for a given off current due to the advantage of gate coupling as discussed earlier. As a result, FinFETs attain large drive currents at reduced power-supply voltages compared to bulk MOSFETs.

11.2.2 Nonlinear Gate Resistance by Three Dimensional Structure

Unlike the bulk MOSFETs, gate resistance of FinFET exhibits a non-linear relationship with respect to channel width. The FinFET has a 3-D channel (known as fin) is wrapped around by the gate. Figure 11.13 shows the three dimensional structure of FinFET. One can analyze from the figure that the

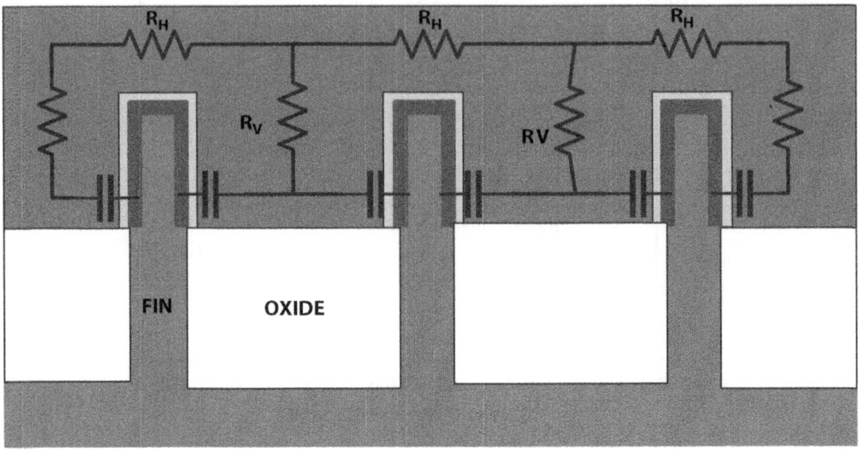

Figure 11.13 Three-dimensional channel of FinFET showing vertical (R_V) and horizontal (R_H) gate resistance components.

gate resistance in FinFET has two components namely, vertical resistance and horizontal resistance [7].

The decomposition of resistance components of FinFET technology is shown in Figure 11.14. The contact resistance R_1 represents the resistance at the edge of the gate material to the metal interface. The combination of R_2 and R_3 is represented by the vertical resistance R_V and the effective R_4 represents open-end resistance. The parallel connection of R_5 and R_6 is represented by the horizontal resistance R_H. The total equivalent gate

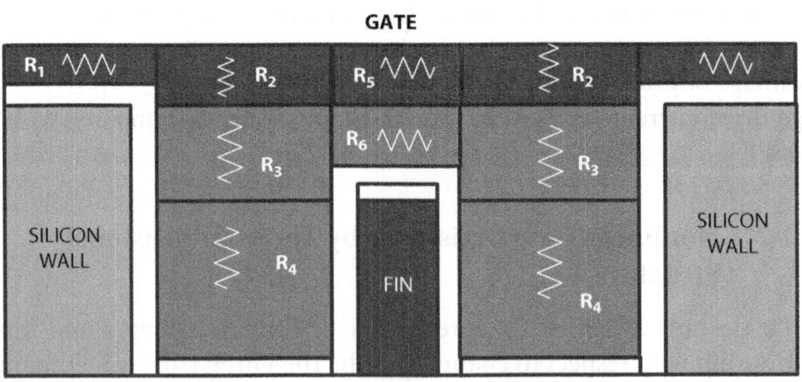

Figure 11.14 Decomposition of resistance components of FinFET technology.

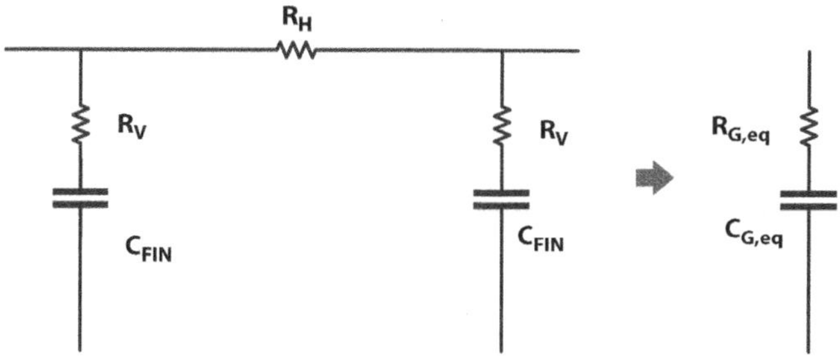

Figure 11.15 Simplified power-equivalent RC network.

resistance of FinFET is denoted by $R_{G,eq}$ or simply R_G and it is calculated by considering the power equivalent network [8] as shown in Figure 11.15.

For the single fin structure, the gate resistance can be calculated as follows:

The gate resistance is given by

$$R_G = R_{CONTACT} + \frac{R_V}{2} + \frac{R_H}{6} \tag{11.1}$$

$$R_{CONTACT} = R_1 \tag{11.2}$$

$$R_V = R_2 + R_3 + \frac{R_4}{3} \tag{11.3}$$

$$R_H = R_5 || R_6 \tag{11.4}$$

Where, $R_{CONTACT}$ is the contact resistance, and R_V & R_H are the vertical and the horizontal resistances. If the number of fin increases, then the gate resistance can be calculated as given by:

$$R_G = R_{CONTACT} + \frac{N}{3}\left(1 - \frac{1}{4N^2}\right)R_H + \left(\frac{1}{N} - \frac{1}{2N^2}\right)R_V \tag{11.5}$$

Where 'N' is number of fins. If 'N' is very large, then the equation for gate resistance can be simplified as

$$R_G \approx R_{CONTACT} + \frac{NR_H}{3} + \frac{R_V}{N} \tag{11.6}$$

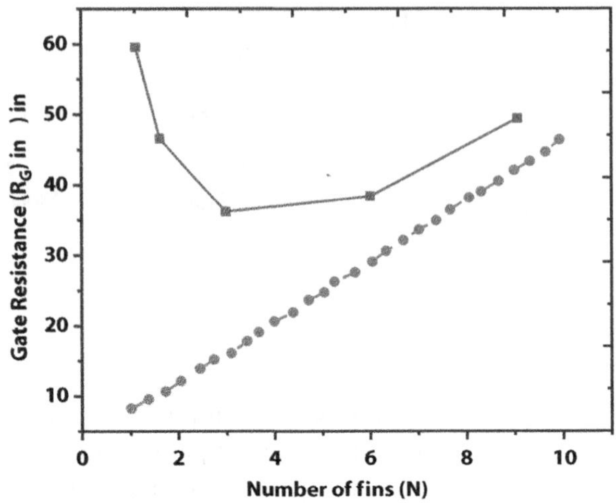

Figure 11.16 Variation of gate resistance with respect to number of fins.

Due to the nonlinear relationship of vertical resistance with the number of fins, gate resistance [9] starts to decrease as 'N' increases until the horizontal resistance begins to dominate. In FinFET, as the width of the channel increases, the R_H becomes stronger, and then the gate resistance shows linear variation with respect to channel width as shown in Figure 11.16.

11.2.3 Self-Heating Effect in FinFETs

In recent years, the self-heating effect (SHEs) [10] has been a major issue in silicon industry because of the continuous downscaling of device dimensions. In planar bulk MOSFETs, the heat produced in the channel dissipates to the substrate easily by lateral spreading. On the other hand, unlike planar MOSFETs, the materials used in FinFETs have poor thermal conductivity; hence the self-heating effect in FinFETs is likely to be more significant than planar MOSFETs.

In CMOS devices, on application of the driving voltage, minority carriers in the channel gains sufficient energy and they tunneled into the cramped volume. Throughout this operation, scattering event of the phonon increases which continues to heat up the silicon substrate. But the heat released in this process does not completely absorbed by the buried oxide, and is accumulated in the channel. Hence, local temperature of the channel increases, which is referred as "self-heating effect".

11.3 Assessment of FinFET Technology for RF/mmWave Applications

Nowadays FinFETs offers superior control over the SCE but their speed of operation is limited to an extent [1] and the scaling of the gate length is insufficient for better RF performance. Therefore, for optimum performance from FinFETs, the understanding of parasitic's major role and their reduction techniques becomes utmost important. In a tri-gate FinFET, a high parasitic capacitance exists and hence, the cutoff frequency (f_T) is less than the planar architecture. Therefore, in spite of good DC characteristics and scaling benefits, one may hesitate to use FinFET for RF/mmWave circuit design because of relatively less peak f_T than the planar technology. In addition to this, the source-drain resistance (R_{SD}) can be reduced with many techniques reported in the literature. The reduction in spacer width causes increase in short channel effect. This would in turn causes higher f_T and lower maximum oscillation frequency (f_{max}) [1], hence better RF performance. It is important to consider the transistor with higher gain and frequency for reliable lifetime operation. The reduction in source-drain resistance without effecting SCE may be achieved by proper selection of epitaxial growth. By using this approach, the speed may increase but at the cost of increase in the parasitic capacitance. The two RF figure of merit, namely f_T and f_{max} to be considered [11] are calculated as under:

$$f_T = \frac{gm}{2\pi\sqrt{C_{gs}^2 + 2C_{gs}C_{gd}}} \times \left\{1 - \frac{C_{gs}+C_{gd}}{C_{gs}^2 + 2C_{gs}C_{gd}} \times [g_{ds}(C_{gs}R_s + C_{gs}R_d + C_{gd}R_d) + g_m(C_{gd}R_d - C_{db}R_s)]\right\} \tag{11.7}$$

$$f_{max} = \frac{f_t}{2}\left\{g_{ds}R_g + 2\pi f_t C_{gd}R_g + \frac{C_{gd}+C_{db}}{C_{gs}+C_{gd}} 2\pi f_t C_{gd}R_g - \frac{C_{db}}{C_{gs}+C_{gd}} 2\pi f_t C_{gs}R_s + \frac{C_{gs}^2 g_{ds}R_d + C_{gs}^2 g_{ds}R_s}{(C_{gs}+C_{gd})^2}\right\}^{-\frac{1}{2}} \tag{11.8}$$

where g_m is transconductance, g_{ds} is the output conductance, C_{gs} & C_{gd} are gate-to-source and gate-to-drain capacitances, R_s & R_d are source and drain extrinsic resistances, C_{db} is drain-to-bulk junction capacitance, and R_g is the gate resistance. In order to obtain the intrinsic characteristics along with g_m and g_{ds}, a 3-D simulation is shown.

11.3.1 RF Performance

With the evaluation of three-dimensional structure, it is observed that the gate resistance has horizontal and vertical resistance components. In the previous discussions, the total gate resistance for multiple fins is calculated. In the total gate material across multiple fins, some of the material between the Fins is not functional, as it does not contribute to the transconductance of the device but still contributes to the parasitic loads ($R_{out,DC}$). The non transconductance contributing section of the gates increases the total gate resistance, the parasitic capacitance [11] between gate-to-drain/source interface ($C_{gd,par}$, $C_{gs,par}$) and the gate to outside channel fin ($C_{g,finr}$). Hence, more gate material between the fins degrades the peak f_T as compared to the planar devices. The recent silicon evidence proves that the peak f_T of FinFET is 20% less than that of planar devices. The vertical portion of gate resistance, reduces the total gate resistance up to a certain number of fins as shown in Figures 11.17 and 11.18. The value of f_{max} is high in case of FinFET which makes it suitable for mmWave circuit design.

From recent literature survey, it has been observed that the peak f_T and peak f_{max} degrades with respect to poly pitch tightening, due to large

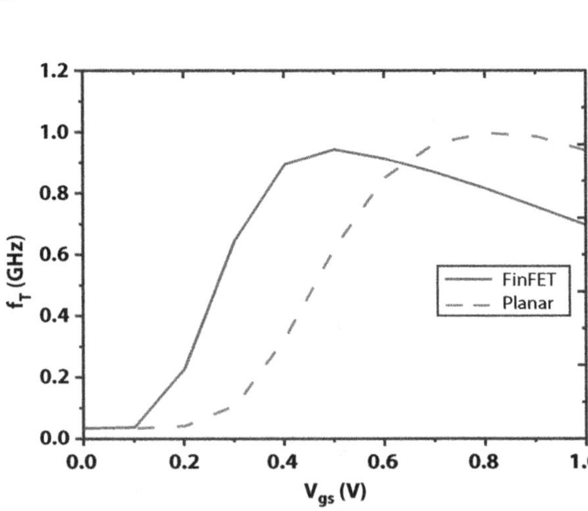

Figure 11.17 f_T variations in FinFET and Planar devices with respect to V_{gs}.

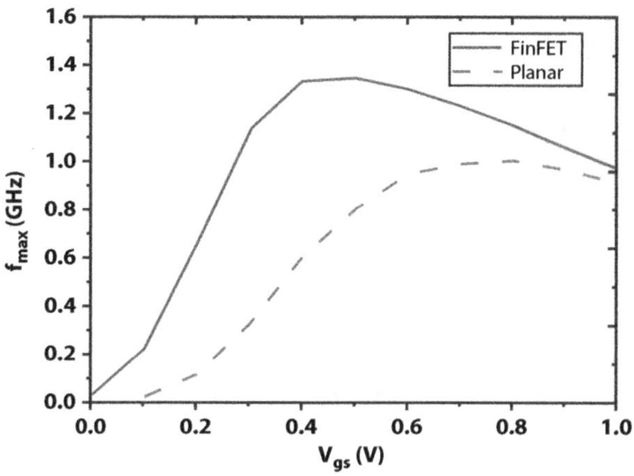

Figure 11.18 f_{max} variations in FinFET and Planar devices with *respect to* V_{gs}.

parasitics. The peak f_T and peak f_{max} in planar devices constantly improved by scaling channel length, increasing the transconductance and decreasing the gate capacitance. In FinFET technology, the enhancements in transconductance are associated with channel length shrinking because of velocity saturation. The complete trend of peak f_T and peak f_{max} has reached 22nm [6, 12] technology peaks for both planar and FinFET technologies as shown in Figures 11.19 and 11.20.

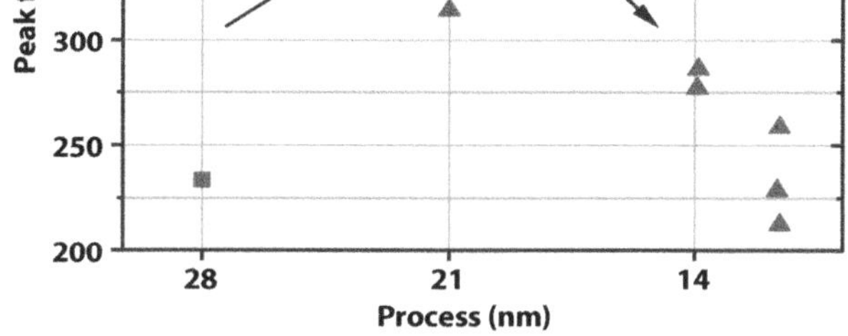

Figure 11.19 Variation in f_T with process technology.

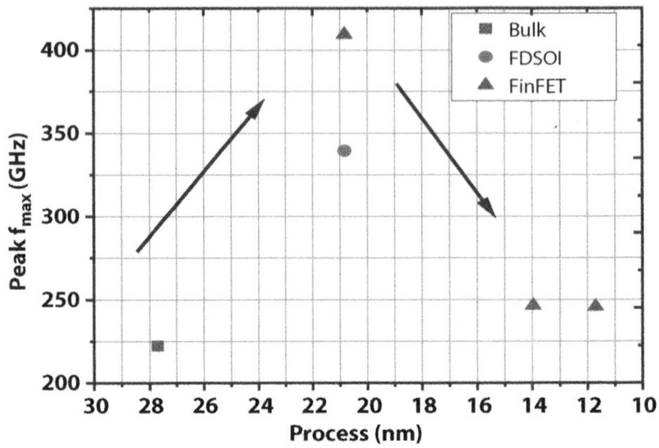

Figure 11.20 Variation in f_{max} with process technology.

11.3.1.1 Parasitic Extraction

1. *Series resistance* (R_S):

The distance between the gate edge and drain (L_{GD}) and Fin width can be adjusted in order to decrease the source-drain resistance R_{SD} only if the technology permits. Several experiments shown that the scaling down of L_{GD} up to 40nm reduces the effective R_{SD} by 25%. It is because the path of current in the extension is decreased and the final effective fin width is increased according to fin curvature.

2. *Gate resistance* (R_G):

Designing of gate resistance (R_G) plays very important role in RF performance as it effects f_{max}. In addition, R_G becomes very significant with

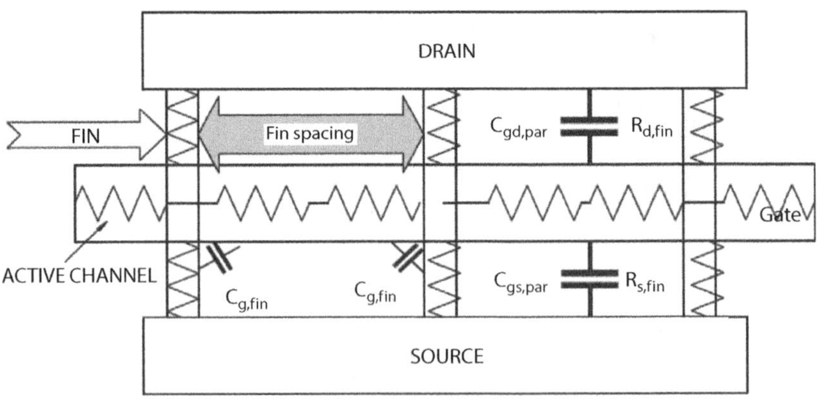

Figure 11.21 FinFET parasitic representation.

downscaling of L_G. Compared with the planar devices, in FinFET devices, the gate is placed over a non-uniform substrate. The gate resistance can be considered by taking a series connection of RC components in transmission lines as shown in Figure 11.21.

3. *Substrate resistance* (R_{sub}):

The substrate resistance(R_{sub}) can be extracted from a conventional off-state equivalent circuit as shown in Figures 11.22 and 11.23 [13]. The extracted R_{sub} of a planar device with a gate length (L_G) of 50 nm, a channel width (W) of 140 nm has notable frequency dependency. Whereas in FinFET, the substrate resistance has less frequency dependency with increase in size of the device.

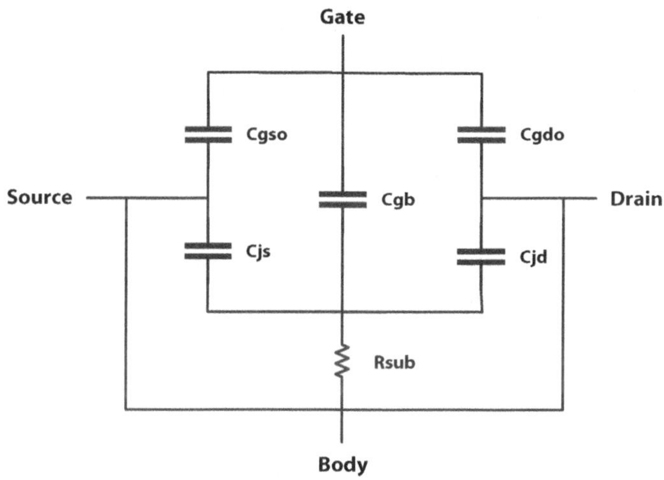

Figure 11.22 Off-state small signal equivalent circuit of MOSFET for RF modelling.

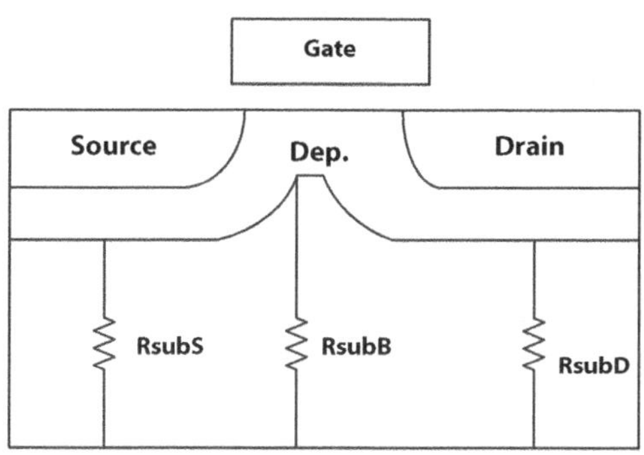

Figure 11.23 Two dimensional cross sectional view of substrate resistances [14].

4. Parasitic capacitance:
A higher value of capacitance is observed in FinFET as compared to a planar transistor. It has been found that the speed of the device may increase if the existing parasitic capacitance of the finger interconnection and back-end-of line (BEOL) is reduced. When the interconnect is de-embedded, the f_T improves by 20%. The drain to source interconnect capacitance can be reduced if the distance between active areas increases and when the layers above metal 1 are suppressed. The scalable model can be implemented by taking only first order parasitic into consideration. The model depicts an increase in C_p with N_{finger}, even when the total width is kept constant, i.e., $N_{finger} \cdot N_{fin}$ = constant. This is proved experimentally by a 15% f_T variation for different folding.

Although the peak f_T is reduced in FinFET technology due to parasitics, yet it offers better performance for RF and mmWave design over the planar. It is worthwhile to note that the peak f_T and f_{max} of planar device required 60% higher V_{gs} than FinFET due to SCEs or DIBL effect.

11.3.2 Noise Performance

With the rapid increase in scaling of CMOS technology, the noise personation in terms of robustness became more important, hence gearing up voltage margin is reduced as transistor size shrinks. So the noise study is important to understand the reliability issues. In addition, $1/f$ can be used as a most reliable diagnostic tool to check the quality of interface between oxide/substrate, gate/oxide and the oxide itself.

For better understanding of noise performance and also to provide the better results for circuit simulation, two physical models are proposed, which explains the $1/f$ noise [14] and speculate a gate bias dependence on noise spectra. The McWhorter theory model demonstrates that $1/f$ noise occurs because of carrier fluctuations (Δn) in the channel due to capturing and releasing of carriers in dielectrics. Whereas, Hooge theory model demonstrates that the noise is generated with mobility fluctuation (Δu), because of carrier separation by photons. To discuss the emergence of flicker-noise, which is still unresolved in the above models, the bias temperature instability (BTI) [15] used as an alternate tool. BTI becomes the critical issue for both NMOS and PMOS devices on application of the high metal gate potential. The positive BTI (PBTI) in NMOS is because of electron trapping, which occurs at high potential. On the other hand, PMOS has worst negative BTI (NBTI) performance.

A two-port noise network is shown in Figure 11.24, where one can reconstruct the noisy two port network to noiseless equivalent two port

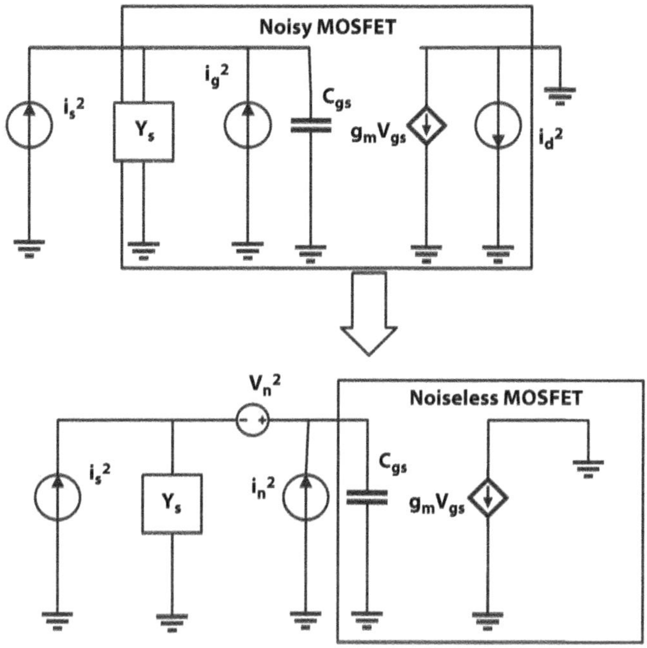

Figure 11.24 Two-port noisy and noiseless models of MOSFET.

network. From the equivalent model, we can derive the noise factor (F), minimum possible noise factor (F_{min}), optimum noise impedance (Z_{opt}), and equivalent noise conductance (G_n) as:

$$F = F_{min} + \frac{G_n}{R_s} | Z_s - Z_{opt} |^2 \tag{11.9}$$

$$F_{min} = 1 + 2\left(\frac{\omega}{w_t}\right) b_3 \Upsilon \tag{11.10}$$

$$Z_{opt} = \frac{1}{\omega C_{gs}}\left(\frac{b_3}{b_2}\right) + j\frac{1}{\omega C_{gs}}\left(\frac{b_1}{b_2}\right) \tag{11.11}$$

$$G_n = \frac{\gamma (\omega C_{gs})^2}{g_m} \tag{11.12}$$

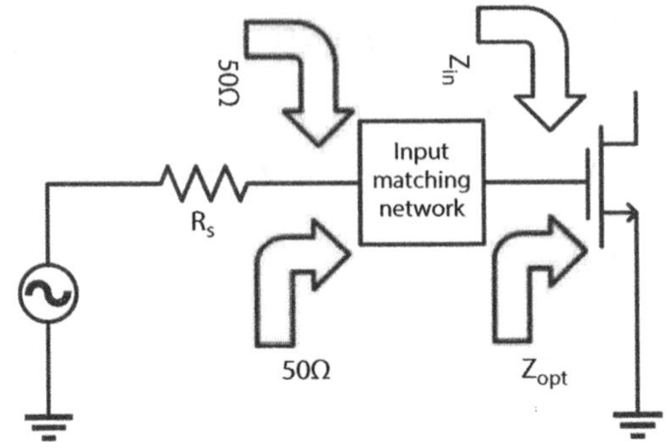

Figure 11.25 f_T and intrinsic gain for the bias condition.

Where

$$b_1 = 1+\Delta|c|, b_2 = 1+2\Delta|c|+\Delta^2, b_3 = \Delta\sqrt{1-|c|^2} \text{ and } \Delta = \sqrt{\frac{\delta}{5\gamma}}$$

Where C is the noise correlation coefficient, γ and δ, are the thermal noise excess factor and thermal noise parameter respectively. The noise correlation coefficient (C) is frequency independent and also decreases as channel length decreases.

From the above equations, it can be observed that overall NF_{min} reduces as unity gain frequency ($wt = 2\pi f_t$) increases and G_n is most sensitive to C_{gs} than w_t. We can conclude that G_n determines the sensitivity of the device to noise mismatch by increasing the amount of mismatch between the source impedance and the optimum noise impedance.

To attain overall improved noise performance *viz.*, lowering NF_{min} and G_n can achieve the lower total noise factor. It has been already determined that the unity gain frequency ωt of FinFET exceeds the planar by more than 50% at reduced bias conditions. Hence, a significant improvement is achieved in thermal noise for the gain at V_{cc} = 1V and 30GH_z as confirmed by the Figure 11.25.

11.3.3 Noise Matching with Gain at the mmWave Frequency

The LNA (low noise amplifier) design suffers due to noise and input mismatch. We know that, there is a mismatch between the conjugate of input impedance $\left(Z_{in}^*\right)$ and optimum noise impedance (Z_{opt}), and Z_{in}^* can be expressed as

$$Z_{in}^{*} = r_g + j\frac{1}{\omega C_{gs}} \tag{11.13}$$

Where r_g is the gate resistance. This mismatch determines that how much noise performance degradation occurs when the input matching network is designed for minimum signal reflection and vice versa. The mismatch between the optimum noise impedance (Z_{opt}) and the input matching impedance (Z_{in}) cannot be neglected for planar devices. Hence, it is a major challenge in LNA design to adjust the matching condition to equalize the two impedances Z_{in}^{*} and Z_{opt} by various circuit techniques, such as inductive source degeneration, etc. To match Z_{in}^{*} with Z_{opt} the relationship to be maintained is:

$$r_g \approx \frac{1}{\omega C_{gs}}\frac{b_3}{b_2} \tag{11.14}$$

$$b_1 \approx b_2 \tag{11.15}$$

The above equations require a low-noise correlation coefficient c and a low-noise gamma factor γ. This γ in FinFET already have been reported but still research is required to validate the low-noise correlation coefficient.

With increase in frequency, the noise matching point is reached where Z_{opt} and Z_{in} become closer to each other. The Z_{opt} and Z_{in}^{*} are conjugate, and so therefore no significant effort is required to design the input matching network.

11.4 Design Process of FinFET for RF/mmWave Performance Optimization

The FinFET oriented logic technology process has been successfully adopted due to improved scalability, low-power requirements and better performance results at very low technology nodes *viz.*, 14-nm and 16-nm technology. With more device counts in LTE supported phones and coming up sub-6GH_z 5G bands, area scaling is key factor to place number of chipsets within specified cellphone area taking power scaling also in consideration.

Table 11.1 Comparison of 14nm FinFET and 28nm Planar FET at different process technology.

Technology parameter	14-nm FinFET (without High-K)		28-nm planar FET (with High-K)	
Device	NFET	PFET	NFET	PFET
L_{gate} (nm)	14	14	30	30
Contact poly pitch (nm)	78	78	126	126
V_{dd} (V)	0.8	0.8	1.05	1.05
Drain saturation current, $I_{dsat}\left(\frac{\mu A}{\mu m}\right)$	1523	1433	670	450
Transconductance, $G_{msat}\left(\frac{\mu S}{\mu m}\right)$	3017	2748	985	395
F_t/F_{max} (GHz)	314/180	285/140	308/159	185/102

RF and analog compatibility of the FinFET technology plays a major role in realizing RF performance along with logic System on Chip (SoC) to enhance the logic power, performance characteristics and area grading. The 14-nm FinFET has clear benefits over 28-nm planar FETs with drive current (I_d) and transconductance (g_m) for a given design specification as shown in Table 11.1 [16].

11.4.1 Cascaded Chain Design Consideration for Wireless System

Wireless system design aims for two primary performance goals: noise and linearity. Obtaining the performance target is one thing, but market competitiveness of the end product requires attention to power dissipation and area. The silicon area of the wireless system is more dominated by passive components, such as an embedded coil. And this coil technology is highly independent of process nodes. Due to design rule complications and excessive parasitics in the device and metal interface within such a less accessible space, the transistor node scaling does not always bring positive impact on silicon area scaling. Hence, the total radio frequency integrated

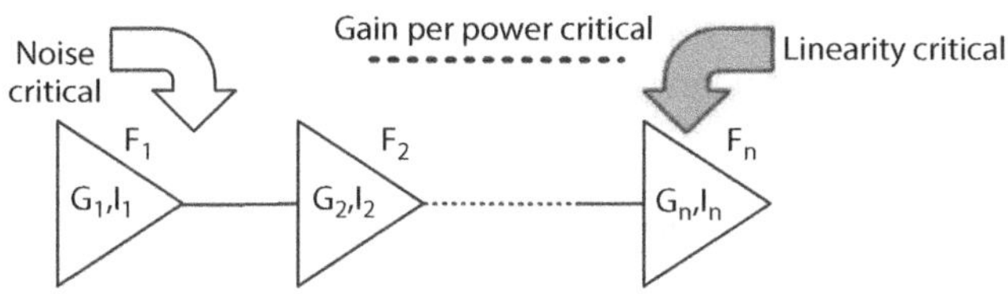

Figure 11.26 Cascaded chain with n-stages.

circuit (RFIC) power dissipation is the major challenging factor among the products in the market.

In wireless system design process, the signal link is optimized by using the cascaded chain shown in Figure 11.26. Noise factor (*F*) and the linearity (*IIP*3) in the cascaded chain are expressed as:

$$F = F_1 + \frac{F_2}{G_1} + \frac{F_3 - 1}{G_1 G_2} + \qquad (11.16)$$

$$\frac{1}{IIP3} = \frac{1}{I_1} + \frac{G_1}{I_2} + \frac{G_1 G_2}{I_3} + \ldots \qquad (11.17)$$

As depicted from equations, the first stage predominates the noise performance (F_{total}), while the last stage dictates the linearity performance ($IIP3_{total}$). The remaining stages will focus more on power efficiency to optimize the overall power dissipation.

11.4.2 Optimization of Noise Figure with G_{max} for LNA Within Self-Heat Limit

With the continuous down scaling of technology, RF performance of CMOS devices has been enhanced accordingly. In the deep-submicron technology, from the device point of view, short-channel MOSFETs exhibits high-frequency figure of merits (FoMs) like cut-off frequency (f_T) and the maximum oscillation frequency (f_{max}) which are more than 100GHz. These FoMs usually do not give any design insights for RF circuit optimization, instead they suggest the flexibility of CMOS devices for operating in RF range. $g_m f_T / I_D$ has been considered as an FoM in case of RF MOSFETs. g_m^2 / I_D has been considered as an FoM in case of analog amplifiers to get gain performance, as it gives power gain per unit dc power consumption.

For LNA design [16], the three basic parameters viz., gain (G), noise figure (F) and power consumption (P) are related as:

$$FoM_{LNA} = \frac{G}{(F-1).P} \tag{11.18}$$

The figure-of-merit may include bandwidth as well, but for this, we need to derive the bias condition for a mmWave LNA design. Mason gain U, a reference device metric for mmWave design methodology is required for G_{max} extension to overcome the F_{max} limitation with neutralization techniques. As discussed earlier, the gate resistance is a nonlinear function of number of fins, so it became an important factor to determine the number of fins for the design under consideration based on the distance between Z_{in}^{*} and Z_{opt} It shows that the peak G_{max} and minimum of Noise figure, NF_{min} can be obtained in a relatively same bias condition.

Each performance factor in LNA's FoM can be analyzed by using small-signal equivalent circuit. Assuming the perfectly matched condition, which is practically suitable at lower target frequency, the signal gain can be given as

$$G = \frac{P_{load}}{P_{in}} = \frac{1}{4}\left(\frac{g_m}{\omega.C_{gs}}\right)^2 \propto g_m^2 \tag{11.19}$$

Here, C_{gs} is constant in the strong inversion region and the signal gain is almost proportional to g_m^2. It is also observed that lower noise figure requires higher g_m, hence high current density is required for planar technology. In FinFET, the overall DC power in the device should be limited or maintained within the certain range to avoid the self-heating effect. By taking the self-heating under consideration, the values for NF_{min} and G_{max} maintained at a constant power limit, called FiSH limit swept for current density, J_D.

The mason gain U is also known for constant power limit after penalized by NF_{min}, written as $U(dB) - NF_{min}(dB)$ This is the modified *FoM* of LNA (FoM_{LNA}) [17]. Taking the FiSH power limit as reference, the DC power term in the original FoM_{LNA} is constant and hence can be removed. With the elimination of DC power term in the equation, the NF_{min} adjusted by Mason gain, $U(dB) - NF_{min}(dB)$ is approximated to the original FoM_{LNA}. The modified FoM_{LNA} is maximum if G_{max} highest bias point and the NF_{min} lowest bias point are closest. For a four-fin device, the maximum FoM_{LNA}

point has been observed while both G_{max} and NF_{min} reach their optimum performance at the closer bias condition.

11.4.3 Gain Per Power Efficiency

Mason has given the unilateral gain, known as Mason gain (U). It describes that maximum stable gain (MSG) is obtained if there is only forward gain in the lossless network. The mason gain (U) is used to measure F_{max} by simply measuring the frequency where the mason gain is unity. It is the absolute measure of F_{max}, where the device operates with the positive power gain if there is any feedback compensation technique available in the device. From Figure 11.27 [8], it is to be noted that for any active device if the frequency increases there will be reduction in its gain due to the increased passive loss in the device network.

As the modern wireless systems used to integrate more wireless standard supporting many frequency bands in a single chip, modern RFIC design emphasizes even more on low-power design.

For a low-power design, one can trade the number of amplifier stages for the total power dissipation by maximizing the gain per stage under the condition that avoids unnecessary power waste for gain as shown in Figure 11.28 [8]. It is estimated that the total number of amplifier stages to achieve a total target gain by dividing over the mason gain of the device by assuming that mason gain is maximum achievable gain per device at a given frequency. Here, the product of I_d/g_m and the total number of amplifier stages

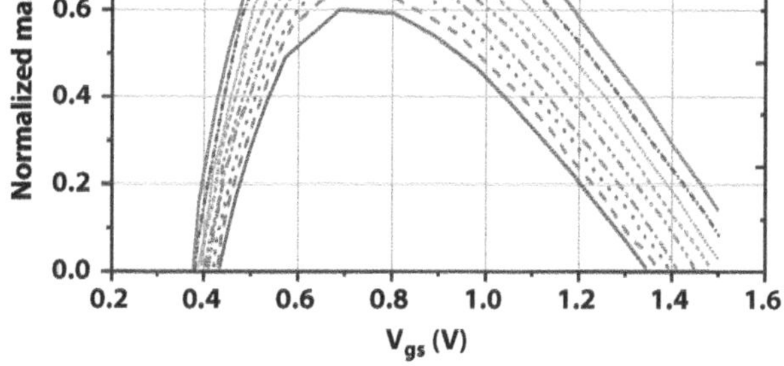

Figure 11.27 Mason gain at different frequencies.

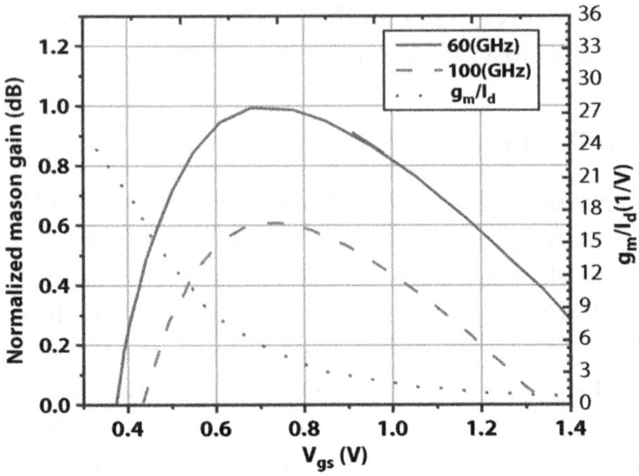

Figure 11.28 Mason gain variation at 60 GHz and 100 GHz with *respect to* V_{gs} (*V*).

is the division of total target gain over mason gain *U*. It is possible to maximize the inverse of the product of I_d/g_m and the number of stages, which can be defined as Gain-Power Figure of merit (FoM_{GP}) given as:

$$FoM_{GP} = U\frac{g_m}{I_d} \tag{11.20}$$

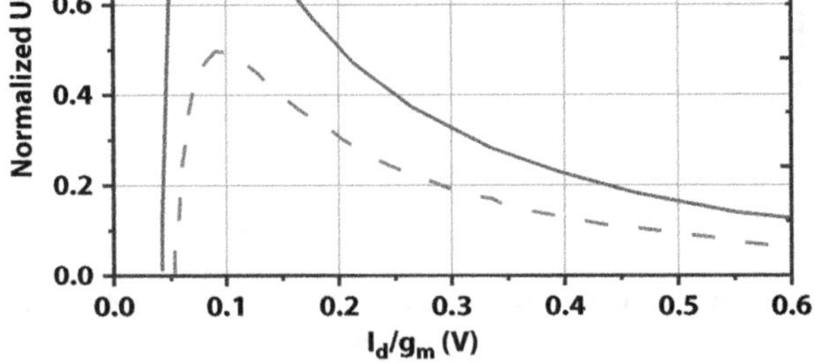

Figure 11.29 Variation in FoM_{GP} with I_d/g_m.

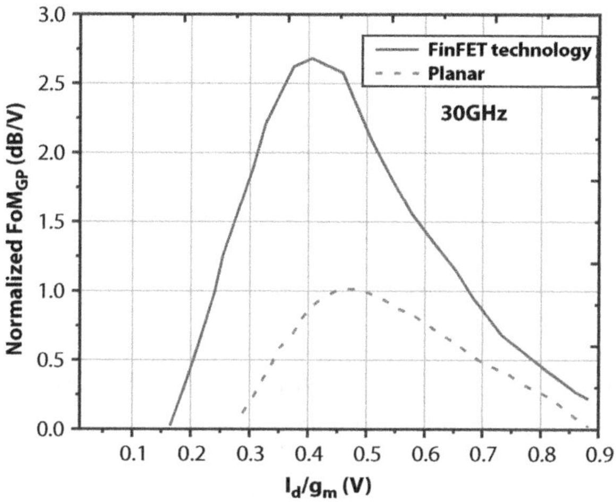

Figure 11.30 Comparison of FoM_{GP} for FinFET and Planar devices at 30GHz with I_d/g_m.

Figure 11.29 shows the FoM_{GP} normalized to peak value at 60 GHz. It has been noticed that FoM_{GP} reaches the peak value at a certain point, which is the optimum bias condition to maximize device gain per power dissipation. The most current planar technology exhibits around a 70–160% improvement in FoM_{GP} over the FinFET technology, which means 40–60% less power dissipation to obtain the same amount of gain as shown in Figure 11.30 [8].

11.4.4 Linearity for Gain and Power Efficiency

Linearity [18] is a critical property for the wireless system configuration represented by *IIP3* or P_{1-dB}. Linearity describes the overall performance degradation by blockers and jammers. Among the various linearity measurement techniques, the double-tone intermodulation distortion is widely used, and is specified by 3rd order intercept point, IP3. This IP3 data can be examined by using either measured or simulated I-V data or double-tone measurement. However, existing reports are mainly for 130nm or older technologies, and depending on first order IP3 theory that relates IP3 to 3rd order derivative of I_{DS} with respect to V_{GS} only. The simple expression of third-order input interception point (*IIP3*) can be given as

$$IIP3 = \frac{1}{6R_s}\frac{(1+(\omega C_{gs}R_s)^2)}{\left|\dfrac{K_{3gm}}{g_m}\right|} \tag{11.21}$$

Where

$$K_{3gm} = \frac{1}{6}\frac{\partial^3 I_{DS}}{\partial^3 V_{GS}}$$

$$g_m = \frac{\partial I_{DS}}{\partial V_{GS}}$$

To increase the linearity, it is required to maximize the ratio called as linearity figure-of-merit (FoM_{LIN}), which approaches to unity when the third-order term K_{3gm} is minimized as given by

$$FoM_{LIN} = \frac{g_m}{g_m + |K_{3gm}|} \tag{11.22}$$

Figure 11.31 depicts that FoM_{LIN} increases as I_d/g_m increases which confirms that the improvement in linearity needs more power dissipation.

If both the gain and linearity improvements are considered, the product of FoM_{GP} and FoM_{Lin} derives another figure-of-merit defined as gain-power-linearity figure-of-merit (FoM_{GPL}), which defines the optimum gain-linearity balanced performance per power dissipation shown in Figure 11.32 [8].

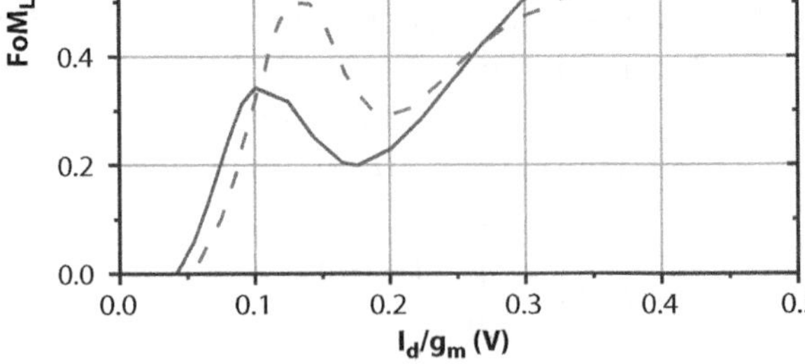

Figure 11.31 Comparison of FoM_{Lin} for FinFET and Planar devices with respect to I_d/g_m.

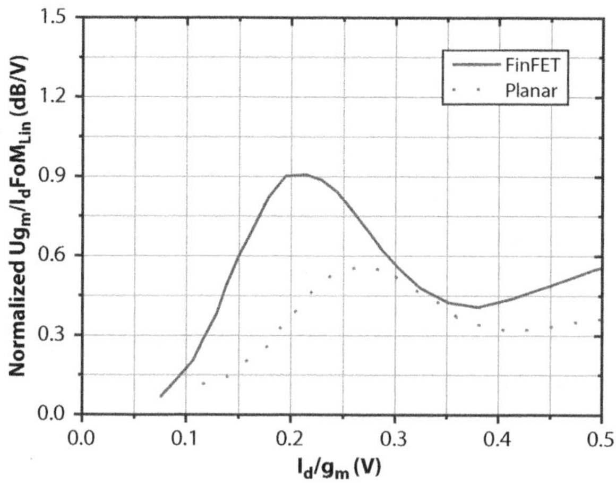

Figure 11.32 Comparison of FoM_{GPL} for FinFET and Planar with respect to I_d/g_m.

This confirms over 80% benefits of FinFET devices over planar devices with less current density requirements.

11.4.5 Neutralization for mmWave Applications

We have compared the FinFET technology against planar using unilateral gain, assuming no feedback network. But, in reality the FinFET offers higher feedback capacitance, C_{gd}, than the planar. This capacitance is a key contributor in limiting the maximum oscillation frequency F_{max}. Hence, it is being advantageous to adopt a technique which can reduce or eliminate

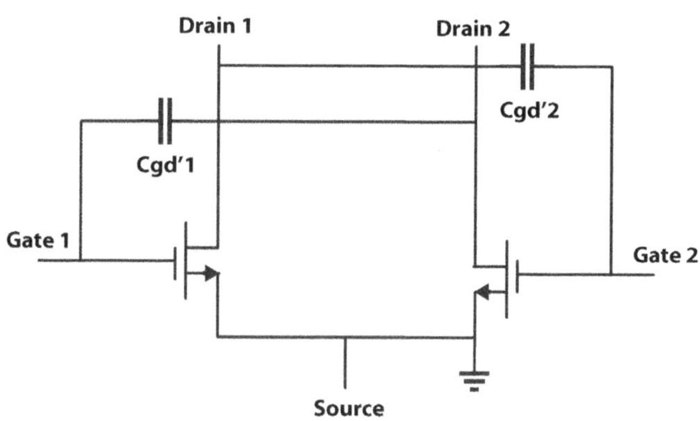

Figure 11.33 Differential neutralization technique.

the feedback capacitance in order to increase the F_{max}. This causes an increase in the gain (G_{max}), generally 4-5dB with the neutralization.

The typical neutralization technique is shown in Figure 11.33. This technique is completely based on cross-coupled capacitor structure to generate the negative C_{gd}, or alternately neutralize the device C_{gd}. This technique consists of differential configuration and provides the neutralization for wide bandwidth range. However, it is evident that the input impedance changes after neutralization. After neutralization, Z_{in}^* and Z_{opt} are no longer close to each other i.e., they move further apart, especially the real part of input impedance. Z_{opt} remains constant before and after neutralization. So, Z_{opt} is less susceptible to the stability issues. These effects need to be considered while applying the neutralization techniques.

References

1. Subramanian, V. *et al.*, Planar bulk MOSFET versus FinFETs: An analog/RF perspective. *IEEE Trans. Electron Devices*, 53, 12, 3071–3079, 2006.
2. Razavi, B., *Design of Analog CMOS Integrated Circuits*, McGraw-Hill, Inc., USA, 2005.
3. Wambacq, P. *et al.*, The potential of FinFETs for analog and RF circuit applications. *IEEE Trans. Circuits Syst. I Regul. Pap.*, 54, 11, 2541–2551, 2007.
4. Colinge, J.-P., *FinFETs and Other Multi-Gate Transistors* (1st. ed.), Springer Publishing Company, Incorporated, 2007.
5. Biswas, K., Sarkar, A., Sarkar, C.K., Impact of fin width scaling on RF/analog performance of junctionless accumulation-mode bulk FinFET. *ACM J. Emerg. Technol. Comput. Syst.*, 12, 4, 1–12, 2016.
6. Nowak, E.J., Aller, I., Ludwig, T., Kim, K., Doubling up. *Eur. Chem. News*, 72, 21, 2000.
7. Lee, H.J. *et al.*, Intel 22nm FinFET (22FFL) process technology for RF and mmWave applications and circuit design optimization for FinFET technology. *Tech. Dig.-Int. Electron Devices Meeting IEDM*, December 2018, pp. 14.1.1–14.1.4, 2019.
8. Hueber, G. and Niknejad, A.M., *Millimeter-Wave Circuits for 5G and Radar*, The Cambridge RF and Microwave Engineering Series, Cambridge, Cambridge University Press, 2019.
9. Scholten, A.J., Smit, G.D.J., Pijper, R.M.T., Tiemeijer, L.F., Mercha, A., Klaassen, D.B.M., FinFET compact modelling for analogue and RF applications. *Tech. Dig.-Int. Electron Devices Meeting IEDM*, vol. 2, pp. 190–193, 2010.
10. Hsieh, E.R. *et al.*, An experimental approach to characterizing the channel local temperature induced by self-heating effect in FinFET. *IEEE J. Electron Devices Soc.*, 6, 866–874, 2018.

11. Sohn, C.W. *et al.*, Device design guidelines for nanoscale FinFETs in RF/analog applications. *IEEE Electron Device Lett.*, 33, 9, 1234–1236, 2012.
12. Tinoco, J.C., Rodriguez, S.S., Martinez-Lopez, A.G., Alvarado, J., Raskin, J.P., Impact of extrinsic capacitances on FinFet RF performance. *IEEE Trans. Microw. Theory Tech.*, 61, 2, 833–840, 2013.
13. Ong, S.N. *et al.*, A 22nm FDSOI technology optimized for RF/mmWave applications. *Dig. Pap.-IEEE Radio Freq. Integr. Circuits Symp*, June 2018, pp. 72–75, 2018.
14. Jung, J.H. and Lee, J.H., Extraction of substrate resistance in bulk FinFETs through RF modeling. *IEEE Microw. Wirel. Compon. Lett.*, 17, 5, 358–360, 2007.
15. Ding, Y.M., Misra, D.D., Srinivasan, P., Flicker noise performance on thick and thin oxide FinFETs. *IEEE Trans. Electron Devices*, 64, 5, 2321–2325, 2017.
16. Singh, J. *et al.*, 14-nm FinFET technology for analog and RF applications. *IEEE Trans. Electron Devices*, 65, 1, 31–37, 2018.
17. Cmos, L.D., Fritsche, D., Tretter, G., Carta, C., Ellinger, F., Millimeter-wave low-noise amplifier design in 28-nm low-power digital CMOS. *IEEE Trans. Microw. Theory Tech.*, 63, 6, 1910–1922, 2015.
18. Song, I. *et al.*, A simple figure of merit of RF MOSFET for low-noise amplifier design. *IEEE Electron Device Lett.*, 29, 12, 1380–1382, 2008.

12

Pre-Distortion: An Effective Solution for Power Amplifier Linearization

Gaurav Bhargava* **and Shubhankar Majumdar**†

Department of Electronics and Communication Engineering, National Institute of Technology (NIT) Meghalaya, Shillong, Meghalaya, India

Abstract

In the broadband transceiver system, Power Amplifier (PA) has maintained a promising place for transmitting the information at wide space. Many circuits of power amplifier in recent times, utilizes Gallium Nitride (GaN) device, which again plays a crucial part in highly efficient PA design. Several class of power amplifier (D, E, F) has high efficiency (100 %) theoretically. The linearity measure has become important factor for characterization of PA. As day by day the methods of modulation are becoming advanced, it eases the process of linearizing an amplifier setup.

This chapter is presented as an update and an overview of Power Amplifier (PA) linearization. In this chapter, we have discussed basic measures of power amplifier nonlinearities. To overcome the effect of those nonlinearities, various types of linearization techniques are discussed. Among them, Analog pre-distortion (APD) finds better space in field of communication system. For verifying the improved linearization, one combined structure of hybrid EFJ power amplifier with APD linearizer block is developed and simulation results are compared with the state of art of variety of linearization schemes.

Keywords: Radio frequency, gallium nitride, power amplifier, analog pre-distortion, linearization

12.1 Introduction

With advancement in the cellular-satellite technology, PA acts as a backbone for providing transmission of data in wide scale. But some linearity

**Corresponding author*: gauravbhargav87@gmail.com

†*Corresponding author*: shub@nitm.ac.in

Sangeeta Singh, Rajeev Kumar Arya, B.C. Sahana and Ajay Kumar Vyas (eds.) RF Circuits For 5G Applications: Designing with mmWave Circuitry, (223–240) © 2023 Scrivener Publishing LLC

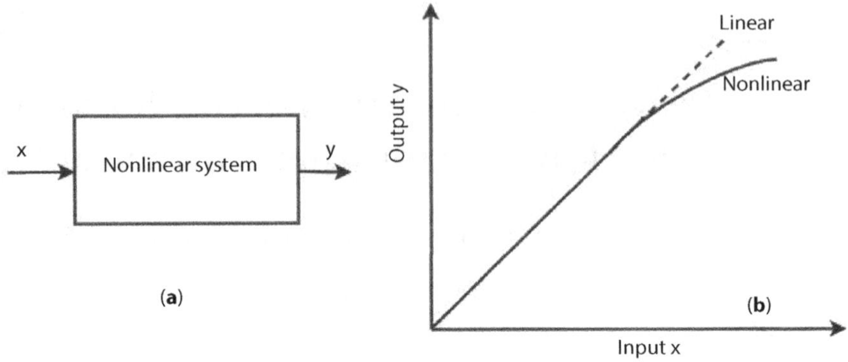

Figure 12.1 Nonlinear system with input x and output y [2].

issues arise while utilizing PA for distant information transfer. Linearization can be performed by many techniques such as feedback, feedforward, and pre-distortion [1]. Linearizer is a basic building block used to linearize the nonlinear system. A nonlinear system has one input x and one output y as shown in Figure 12.1 (a). Generally, response of variation in the output of a linear system follows the same geometry as that of the input variation [2]. This implies that the Gain is constant and independent on the actual input level as shown in Figure 12.1 (b).

But, in case of nonlinear system, the response of output gets differ w.r.t the variation in input. This is due to the effects of nonlinearities introduced in the system. The nonlinearity of a system can be modelled in various ways. The easiest would be the use of a polynomial approximation of Volterra series as given in Equation (12.1). Where, a_1 is the desired linear coefficients and other power series coefficients are deviated from the ideal constant Gain.

$$y = a_1 x + a_2 x^2 + a_3 x^3 + a_4 x^4 + a_5 x^5 \tag{12.1}$$

12.2 Standard Measures of Nonlinearity of Power Amplifier

Linearity term can be stated as the capability of a power amplifier to achieve similar changes in the output response as per the variation in the input. But these similar variations do not exist for a long time as power amplifier starts saturating at a particular level of input. Linearity measures that are generally used for the characterization of PA are as follows.

12.2.1 Gain Compression Point (1 dB)

Gain compression point at 1 dB is the measure of a particular value of the output power of the nonlinear PA at which the Gain of the amplifier differs by 1-dB from Gain of the ideal amplifier. This deviation of the gain value can be stated as gain compression. The x-axis and y-axis coordinates of transfer characteristics are denoted by IP_{1dB} and OP_{1dB}, respectively, as shown in Figure 12.2. Due to the existence of nonlinearities in the PA response, the output power level gets compressed before reaching the saturation point [3].

12.2.2 Harmonic and Intermodulation Distortion (IMD)

The output voltage from an amplifier is given by the following power series equation:

$$V_{out} = a_1 V_{in} + a_2 V_{in}^2 + a_3 V_{in}^3 + \ldots\ldots\ldots\ldots \quad (12.2)$$

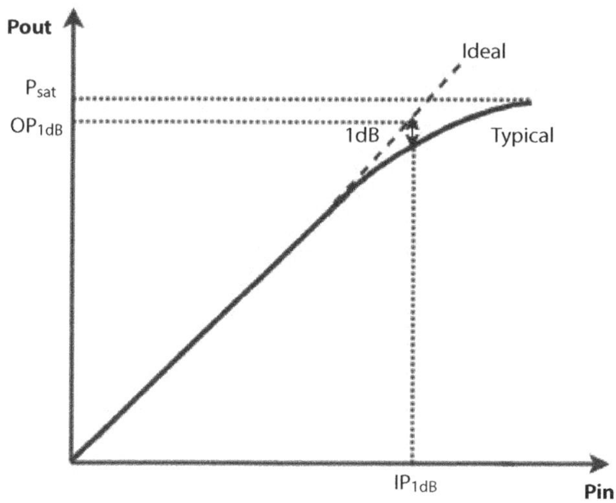

Figure 12.2 Power transfer characteristics of power amplifier [3].

Let's assume the 2-tone input signal V_{in} at frequencies ω_1, ω_2.

$$V_{in} = V\,(\text{Cos}\omega_1 t + \text{Cos}\omega_2 t) \quad (12.3)$$

On Substituting (12.3) in (12.2), following equation is obtained

$$V_{out} = a_1 V\,(\text{Cos}\omega_1 t + \text{Cos}\omega_2 t) + a_2 V^2(\text{Cos}\omega_1 t + \text{Cos}\omega_2 t)^2 + a_3 V^3(\text{Cos}\omega_1 t + \text{Cos}\omega_2 t)^3 + \ldots\ldots\ldots\ldots \quad (12.4)$$

After simplifying the Equation (12.4) we can get the following products occur at the various frequencies:

- DC Term: $a_2 V^2$
- Fundamental Tone (ω_1, ω_2): $a_1 V$
- Second-order Terms ($\omega_1 + \omega_2$, $\omega_2 - \omega_1$): $a_2 V^2$
- Third-order Terms ($2\omega_2 - \omega_1$, $2\omega_1 - \omega_2$, $2\omega_2 + \omega_1$, $2\omega_1 + \omega_2$): $\frac{3}{4} a_3 V^3$

Third order Intermodulation Distortion (IMD3) is defined as the ratio of magnitude of the third-order term and magnitude of the first-order term as given in Equation (12.5).

$$\text{IMD3} = (\tfrac{3}{4}\, a_3 V^3)/\,(a_1 V) = \; 3\, a_3 V^2/4a_1 \quad (12.5)$$

The magnitude of the IMD3 terms as shown in Figure 12.3 can be reduced when the value of the magnitude at frequencies $2\omega_2 - \omega_1$ and $2\omega_1 - \omega_2$ is lowered [3, 4].

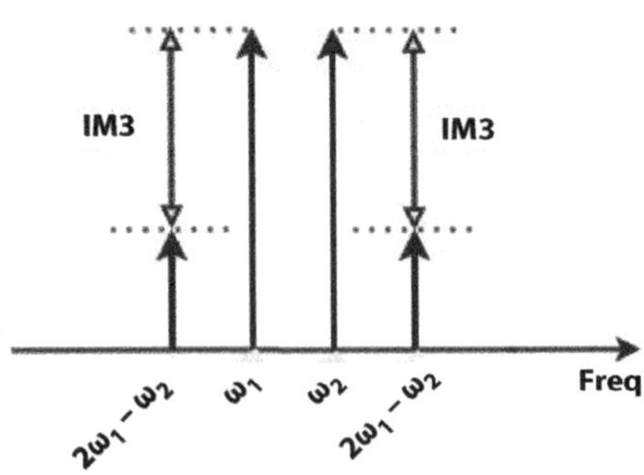

Figure 12.3 IMD3 response with frequency [4].

12.2.3 Third-Order Intercept Point (TOI)

Generally, for every 1 dB increase of fundamental signal (with slope = 1), there is a 3 dB increment in the value IMD3 signal (with slope = 3), as shown in Figure 12.4. After a certain range of values, it is observed that there is an existence of compression in the levels of both the fundamental and IMD3 signals; moreover, on extending the slope lines of both the signals to an intercept point named Third-order intercept (TOI). At this intercept point, the output and input power value are termed third order output intercept (OIP3) and third order input intercept (IIP3), respectively.

A simplified relation of OIP3 and the IMD3 can be expressed mathematically by Equation (12.6),

$$IMD3 = 2(P_{out} - OIP3) = 2(P_{in} - IIP3) \tag{12.6}$$

12.2.4 AM/AM and AM/PM Distortion

The measure of nonlinearity of power amplifier is generally defined by amplitude and phase of the input and output signal. When the nonlinear power amplifier is excited by giving some input, there exists a change in the value of amplitude and phase of the output signal of PA. Amplitude Modulation-Amplitude Modulation (AM/AM) and Amplitude Modulation-Phase Modulation (AM/PM) distortion can be defined as the measure of variation in the amplitude/phase of the output signal due

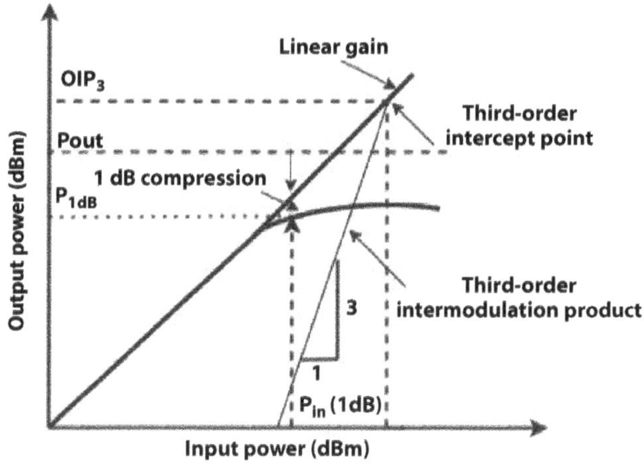

Figure 12.4 Third-order intercept with IMD3 [4].

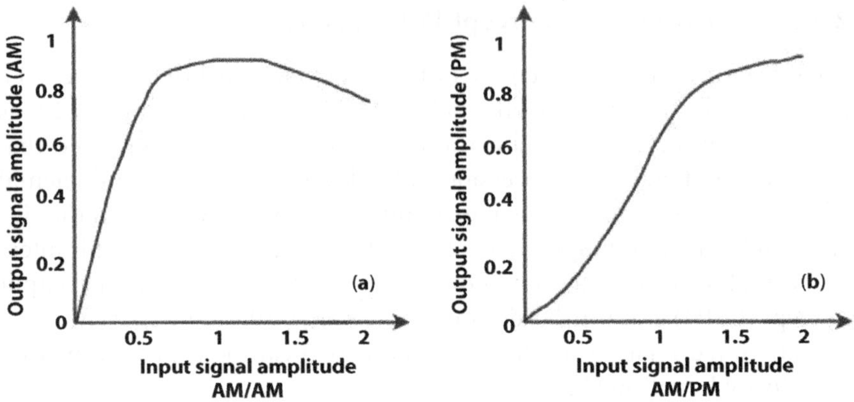

Figure 12.5 Graphical representation of (a) AM/AM and (b) AM/PM distortion [4].

to variation in the amplitude of input signal, respectively as shown in Figure 12.5. The mathematical expression is given by Equation (12.7) and Equation (12.8).

$$V_i = \alpha\ (t) \cos\ (\omega_0\, t + \theta\ (t)) \tag{12.7}$$

$$V_o = G\ (\alpha\ (t)) \cos\ (\omega_0\, t + \theta\ (t) + \phi\ (\alpha\ (t)) \tag{12.8}$$

Where $\alpha(t)$ and $\theta(t)$ are the amplitude and phase of the input signal, respectively; $G(.)$ and $\phi(.)$ are the amplitude and phase characteristic functions of the PA output, respectively.

Existence of distortion in amplitude or phase of PA involves some non-linear effects such as compressed Gain, saturated power, and occurrence of phase shift in output response of PA.

12.2.5 Adjacent Channel Power Ratio (ACPR)

Due to the existence of non-linearities in power amplifier, spreading of spectrum on each side occurs adjacent to the main channel, as shown in Figure 12.6. Another linearity metric called Adjacent Channel Power Ratio (ACPR) can be measured by taking the ratio of mean power of main channel to the mean power of the adjacent channel as given by Equation (12.9).

$$\text{ACPR} = \frac{\text{mean power of the main channel (1)}}{\text{mean power of the adjacent channel (2 or 3)}} \tag{12.9}$$

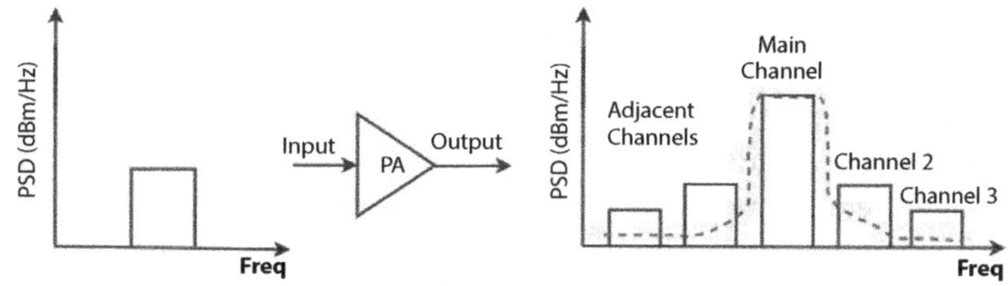

Figure 12.6 ACPR response for power amplifier [5].

12.2.6 Error Vector Magnitude (EVM)

Error vector magnitude can be measured by taking ratio of the root mean square power (RMSP) of error vector signal to the reference signal [5]. Figure 12.7 shows the error vector representation of EVM. The mathematical representation of EVM is given by Equation (12.10).

$$\text{EVM(dB)} = 10\log_{10}\left(\frac{\text{RMSP of error vector signal}}{\text{RMSP of reference signal}}\right) \quad (12.10)$$

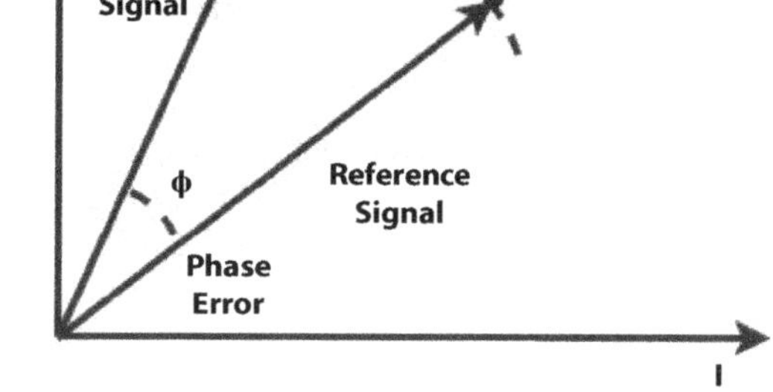

Figure 12.7 Error vector diagram [5].

12.3 What is Linearization?

Linearization is the method to find the optimum way to suppress the non-linearity effects of Power Amplifier. It is the way to linearize PA by pre-applying different types of analytical methods on the actual input (before linearization) to the PA. After the application of these methods, the output of the linearizer is fed as a modified input to the PA. Some linearization techniques are discussed below.

12.3.1 Feed Forward Linearization

The simplified block diagram of feed forward linearization is shown in Figure 12.8. The applied input Vin is firstly split by power divider and fed through two ways: the upper path goes to the main amplifier and the lower path to the auxiliary amplifier. After the amplification of signal through the main amplifier in upper path, the generated output V_s, which is equal to the main amplified signal along with the IMD terms. In lower path, after subtracting the main signal from the amplified signal, the resultant signal V_e will contain only the remaining IMD terms. This resultant signal is then applied to the error amplifier for amplification. The output of error amplifier V_{eo} is then subtracted from the main amplified signal V_s. The final output signal V_o is generated, which contains the amplified signal of undistorted nature [6].

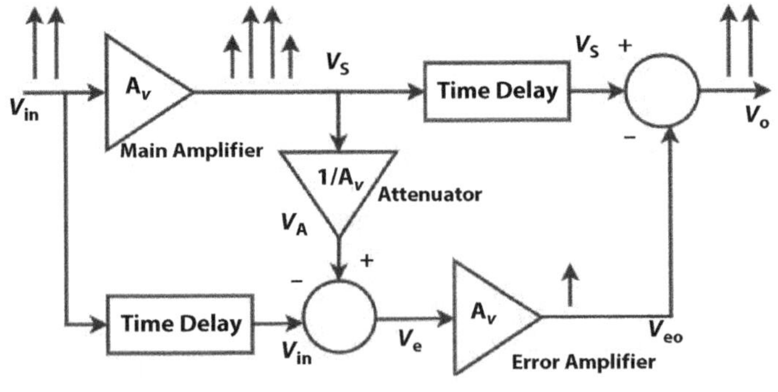

Figure 12.8 Feed forward linearization [6].

12.3.2 Feedback Linearization

The feedback linearization is another type of technique widely used for linearizing the PA. In this type of linearization, output is fed again back to the input. In this feedback process, output and input signal are out of phase. When the output of PA V_f is subtracted with the input signal V_{in}, an error is produced as shown in Figure 12.9. The error at the output V_{PA} is fed back to the input of PA to minimize the resultant difference between them. Generally, negative feedback is used for many cellular communication systems [7].

12.3.3 Pre-Distortion Linearization

Pre-distortion linearization method is one of the most studied techniques used for power amplifier linearization. In this technique, one block of non-linear behavior is added before the nonlinear power amplifier for linearization, as shown in Figure 12.10. This block is called Pre-distorter, which has the gain transfer characteristic opposite of PA. The nature of added pre-distortion block is either analog or digital depending upon the purpose of application for which it is applied [8]. The process of pre-distortion is mathematically simplified as given in Equations (12.11–12.13).

$$x_{PD} = G(x) \tag{12.11}$$

$$Y = k\,F(x_{PD}) \tag{12.12}$$

$$Y = k\,F(G(x)) \tag{12.13}$$

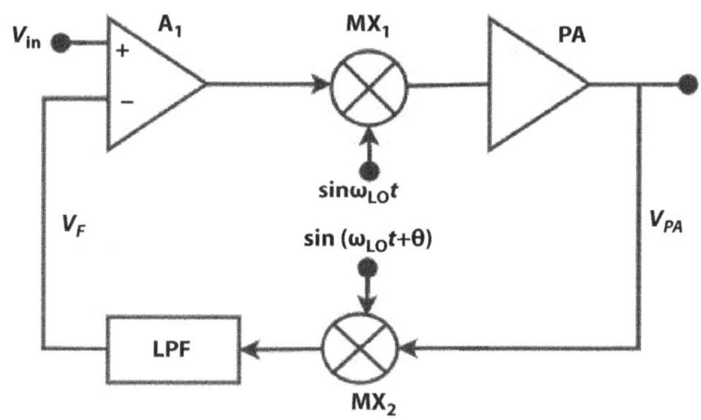

Figure 12.9 Feedback linearization [7].

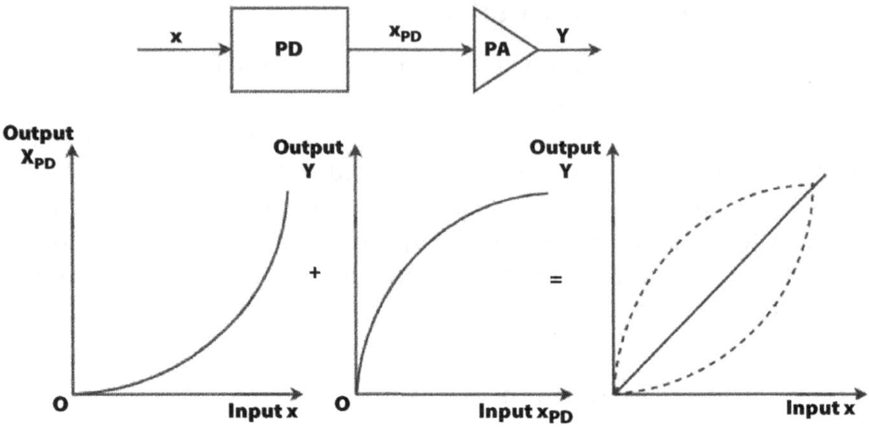

Figure 12.10 Graphical representation of pre-distortion [8].

Here, k is constant; G is transfer of pre-distorter which should be an inverse characteristic of F for linear PA.

(i) Analog Pre-Distortion (APD)

The simplified structure of analog pre-distortion [9] is shown in Figure 12.11. It can be designed by using power splitter (WPD), power combiner (WPC), combination of diodes with resistor and capacitor, Phase shifters (PS), and Directional coupler. Firstly, the applied Radio Frequency (RF) input is split in two paths.

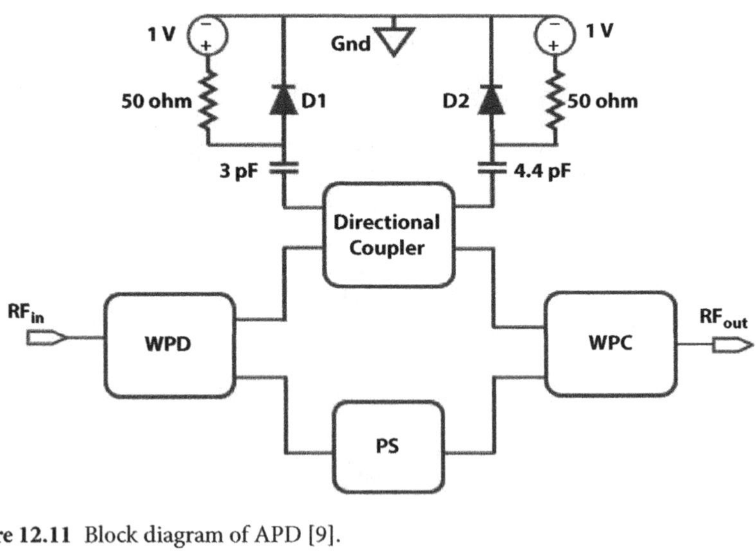

Figure 12.11 Block diagram of APD [9].

The upper path is nonlinear which consist of combination of diodes connected with directional coupler. The lower path is linear contains phase shifter (combination of passive elements) which is used to cancel out the effects of nonlinearity generated at the upper path when signals from upper and lower path are combined through power combiner [10].

(ii) Digital Pre-Distortion (DPD)

With the development of advanced digital processing systems, the requirement of DPD model is highly in demand. Now, it has become a popular linearization technique for power amplifier design. As compared to analog, sophisticated algorithm can be developed in digital scenario. Figure 12.12 shows block diagram of a digital pre-distortion architecture [11].

The RF output from the nonlinear PA is down converted into the baseband signal form, and then it is given to analog to digital converter (ADC). After getting the output in digital form, the estimation of pre-distortion parameters is done. Then by utilizing those parameters, baseband input is pre-distorted. The output of pre-distorter block is given to digital to analog converter (DAC) whose output is up-converted and given to PA for linearization.

A short comparison of merits and demerits of linearization methods utilized for characterization of PA linearity is discussed in Table 12.1.

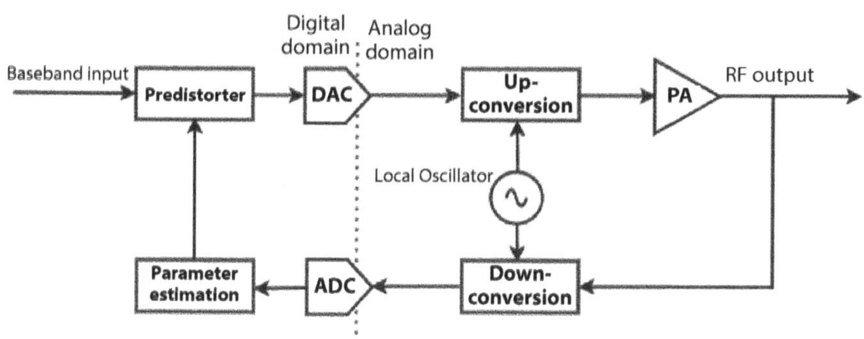

Figure 12.12 Block diagram of DPD [11].

Table 12.1 Comparison of various linearization methods.

Linearization technology	Advantages	Disadvantages
Feedforward	Relatively great improvement in linearity, relatively high bandwidth	Additional power amplifier, dependency on stable operational conditions
Negative feedback	High accuracy, low price	Existence of bandwidth limitation, less stability, limited applications
Analog Pre-distortion	Easy to design, more suitable for higher operating range applications	Dependency on stable operational conditions, limited Linearization capability
Digital Pre-distortion	Relatively easy adaptation implementation, versatile architecture	High energy consumption, Expensive converters

12.4 Example of Analog Pre-Distortion-Based Class EFJ Power Amplifier

In order to verify the effectiveness of the predistortion to linearize a power amplifier, a hybrid Class-EFJ PA [12] is developed on substrate (dielectric(ε_r) = 3.66, Height = 30 mil) using CG2H40010F Cree GaN High Electron Mobility Transistor (HEMT) device at operating frequency of 3 GHz. The drain side voltage (V_{DS}) and gate side voltage (V_{GS}) is 28 Volts and −2.7 Volts, respectively. This type of hybrid power amplifier [13] is structured by using transmission line arrangement in order to produce orthogonal coupling.

As discussed in section 12.3.3 (i), The APD is type of linearizer block which can be used in cascade with PA. The combined network consists of schematic of EFJ PA precedes with APD as shown in Figure 12.13.

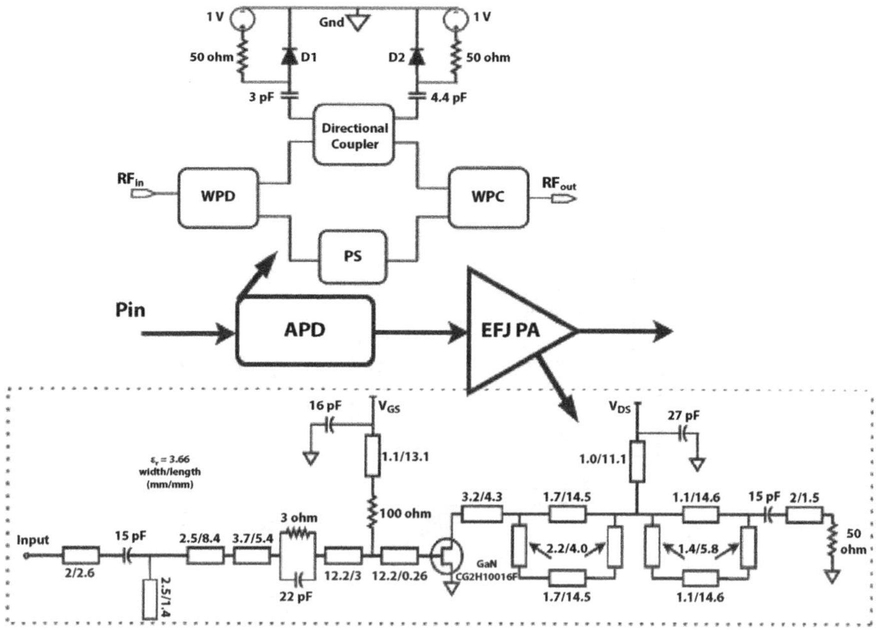

Figure 12.13 Schematic of APD with EFJ PA.

When two-tone harmonic balance analysis is performed, the simulated output of third and fifth harmonics with and without APD are shown in Figure 12.14 (a). For ACP analysis, 64-Quadrature amplitude modulation (QAM) Orthogonal frequency division multiplexing (OFDM) signal is applied to the input of combined network (EFJ PA with APD), simulated output spectrum is achieved, as shown in Figure 12.14 (b).

After performing the simulation of PA with APD, it is clearly noticed that the linearity measures are getting better, and the power amplifier is close to becoming a linear one (results comparison is shown in Table 12.2). Table 12.3 gives the comparative analysis with state of art linearization techniques.

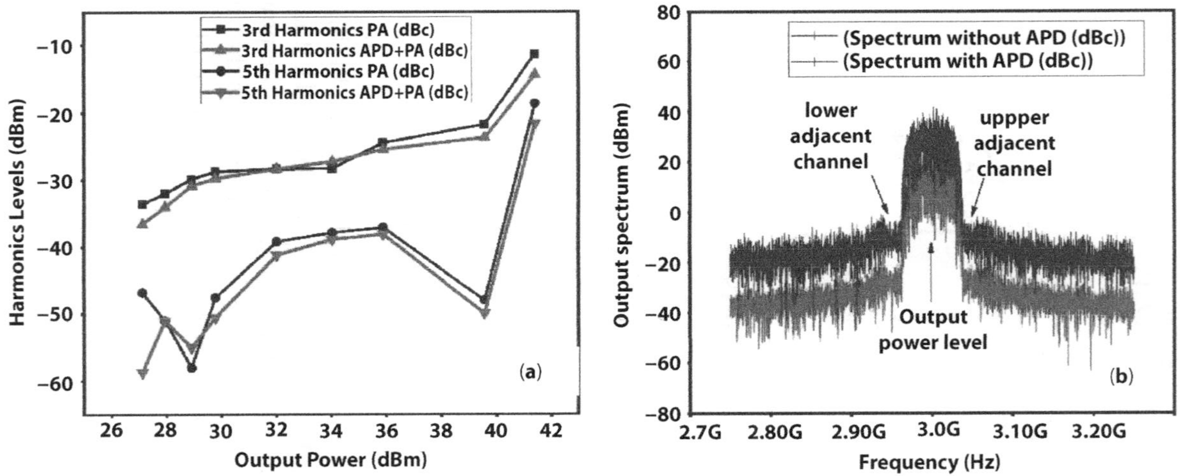

Figure 12.14 Simulation results of EFJ PA with APD (a) third & fifth order harmonics (b) output spectrum vs frequency.

Table 12.2 Comparison of results with and without APD.

Linearity measures	EFJ PA	EFJ PA with APD
Third order harmonics (dBc)	-27.1	-30.8
Fifth order harmonics (dBc)	-36.2	-39.4
ACP lower (dBc)	23.2	39.3
ACP upper (dBc)	23.6	40.1
Adjacent lower	28	15.1
Adjacent upper	27.7	14.9

Table 12.3 Summary of IMD3 improvement with state of art linearization methods.

Ref	Operating frequency (GHz)	Device technology	IMD3 improv. (dBc)	Linearization scheme
[14]	2.14	LDMOS	6	APD
[15]	2.4	GaN HEMT	10	APD
[16]	1.9	GaN HEMT	4	Diode predistortion
[17]	8-10	GaN HEMT	4	Capacitance compensation
[18]	0.6	LDMOS	4.9	Pre-distortion
[19]	2.16	LDMOS	3.88	RF Pre-distortion
This work	3	GaN HEMT	4	APD

Conclusion and Future Scope

In this chapter, the bunch of six linearity measures (1 dB compression, ACPR, EVM, TOI, IMD3, AM/AM, and AM/PM distortion) is discussed with some mathematical relations. The promising techniques (feedback, feedforward, and analog/digital pre-distortion) have drawn greater attention of many researchers. Each technique has certain limitations in utilization specifically for particular application. Also, for practical observation of PA linearization, one example of PA with APD is discussed with simulated output response. In the future, other devices like Gallium arsenide [20] can be utilized for PA design.

References

1. Kenington, P.B., *High Linearity RF Amplifier Design*, 1st ed., Artech House, Norwood, MA, USA, 2000.
2. Chuen, W. and Neo, E., *Highly Efficient and Highly Linear RF Power Amplifiers*, Delft University of Technology, Netherlands, 2010.
3. Fayed, K.A., Zaghloul, A., II, Davis, W.A., Riad, S.M., Ezzeddine, A.K., *The Design of Linearized Power Amplifier for Wireless*, Virginia Tech, Falls Church, VA, 2009.
4. Nan, J. and Gao, M., *Nonlinear Modeling Analysis and Predistortion Algorithm Research of Radio Frequency Power Amplifiers*, CRC Press, Boca Raton, Florida, 2021.
5. Saini, K., Raman, S., Yi, Y., Hudait, M.K., Nazhandali, L., Guido, L.J., *Linearity Enhancement of High Power GaN HEMT Amplifier Circuits*, Virginia Tech, Blacksburg, Virginia, 2019.
6. Pothecary, N., *Feedforward Linear Power Amplifiers*, Artech House Publishers, Norwood, MA, USA, 1999.
7. Borel, A., Barzdėnas, V., Vasjanov, A., Linearization as a solution for power amplifier imperfections: A review of methods. *Electronics (Basel)*, 10, 9, 1073, May 2021.
8. Singh, S. and Malik, J., Review of efficiency enhancement techniques and linearization techniques for power amplifier. *Int. J. Circuit Theory Appl.*, 49, 3, 762–777, Mar. 2021.
9. Bhargava G. and Majumdar, S., Design of novel rat-race coupler based analog pre-distortion circuit for 5G applications, *2022 International Conference for Advancement in Technology (ICONAT)*, pp. 1–4, 2022.
10. Kumar, A. and Rawat, M., A novel analog predistortion linearization technique for RF-in RF-out power amplifier, in: *2019 IEEE MTT-S International Microwave and RF Conference, IMARC 2019*, pp. 1–5, Dec. 2019.
11. Wood, J., Digital pre-distortion of RF power amplifiers, in: *Proceedings of the 2017 IEEE Topical Conference on Power Amplifiers for Wireless and Radio Applications, PAWR 2017*, pp. 1–3, Mar. 2017.
12. Zhang, Z., Cheng, Z., Ke, H., Liu, G., A broadband high-efficiency power amplifier by using branch line coupler. *IEEE Microw. Wirel. Compon. Lett.*, 30, 9, 880–883, Sep. 2020.
13. Bhargava G. and Majumdar, S., Highly linearized GaN HEMT based class E/F3 power amplifier, *2021 Devices for Integrated Circuit (DevIC)*, pp. 564–568, 2021.

14. Lee, Y.S., Lee, M.W., Jung, S.W., Jeong, Y.H., *Analog Predistortion Power Amplifier Using IMD Sweet Spots for WCDMA Applications*, 2007.
15. Cai, Q., Che, W., Kaixue, M., A linear GaN power amplifier using novel transistor based analog predistortion method, in: *2016 IEEE MTT-S International Microwave Workshop Series on Advanced Materials and Processes for RF and THz Applications, IMWS-AMP 2016-Proceeding*, pp. 1–4, Oct. 2016.
16. Akihiro, A., Yoichiro, T., Tsuyoshi, Y., Ryo, I., Kazuhiko, H., A high-efficiency class-f GaN HEMT power amplifier with a diode predistortion linearizer, in: *Asia Pacific Microwave Conference 2008*, Hong Kong and Macau, China, pp. 1–4, 2008.
17. Xie, S. *et al.*, High linearity GaN HEMT power amplifier with pre-linearization gate diode, in: *Proceedings. IEEE Lester Eastman Conference on High Performance Devices, 2004*, pp. 223–228, 2004.
18. Seo, M. *et al.*, Ultrabroadband linear power amplifier using a frequency-selective analog predistorter. *IEEE Trans. Circuits Syst. II Express Br.*, 58, 5, 264–268, May 2011.
19. Lim, K.H. *et al.*, A 60-W multicarrier WCDMA power amplifier using an RF predistorter. *IEEE Trans. Circuits Syst. II Express Br.*, 56, 4, 265–269, 2009.
20. Bhargava, G., Vadalà, V., Majumdar, S., Crupi, G., Auto-encoder based hybrid machine learning model for microwave scaled GaAs pHEMT devices. *Int. J. RF Microw. Comput.-Aided Eng.*, 32, 2022, https://doi.org/10.1002/mmce.23339.

13

Design of Control Circuit for Mitigation of Shadow Effect in Solar Photovoltaic System

Dhvanit Bhavsar, Shubham Bhatt, Siddhi Vinayak Pandey and Alok Kumar Singh*

Department of Electrical Engineering, Adani Institute of Infrastructure Engineering, Ahmedabad, Gujarat, India

Abstract

The quadrant shifting from generation to power dissipation region due to shadow effects in solar photovoltaic affects the performance at module level. Furthermore, the decreased energy in any particular module creates a mismatch at the string level, and hence the overall system performance gets affected heavily. Techniques such as MPPT (Maximum Power Point Tracking), Bypass Diode, Bridge Linked (BL), Total Cross Tied (TCT), etc., are used to mitigate the shading effects up to some extent. Still, none of them are much effective in tackling this issue properly. In this manuscript, the module is divided into 18 small groups containing four cells each to mitigate the shading issue by bypassing the current at individual group level through help of MOSFET (Metal Oxide Semiconductor Field Effect Transistor) switches. If any group/s gets shaded, the controlled switches automatically perform an action based on the gate pulse given through the controller to bypass the shaded group from the circuit. The experimental results cause 27.78% and 55.56% more energy generation compared with conventional module architecture with 72 cells through the activation of one and two bypass diode respectively.

***Keywords*:** Partial shading, photovoltaics, PV modules, losses, bypass

**Corresponding author*: alok.singh@adaniuni.ac.in

Sangeeta Singh, Rajeev Kumar Arya, B.C. Sahana and Ajay Kumar Vyas (eds.) RF Circuits For 5G Applications: Designing with mmWave Circuitry, (241–266) © 2023 Scrivener Publishing LLC

13.1 Introduction

Energy is the ultimate source for work that requires to be done for performing any specific task. There are different forms of energy available: Electrical Energy, Mechanical Energy, Kinetic Energy, Thermal Energy, etc. [1, 2]. There is a massive usage of fuel in producing electrical energy; this leads to the consumption of natural resources for filling the energy need [3].

The renewable sector is easy to operate and has fulfilled the energy requirement for many years. The renewable energy sector includes solar energy, wind energy, hydro-energy, etc. And among them, the use of solar photovoltaics to fulfill the energy need is increasing drastically [4, 5]. Solar energy is the most commercial and user-friendly technique to obtain electrical power from sunrays. Electricity is generated from solar cells by the photovoltaic effect due to the generation of electron and hole pairs in a semiconductor material. Furthermore, these generated electrons are further collected with the help of a current collector (metallic strip) placed on semiconductor material.

Till now, many types of solar cells with different materials are made, such as Crystalline Solar cells (containing - Monocrystalline and Polycrystalline based solar cells), Thin Film Solar Cell (containing - Amorphous/Microcrystalline or Micromorph Tandem Silicon Solar cells, Dye-Sensitized Solar Cell, Organic Solar Cell, Cadmium Telluride Solar Cell, and CIS/CIGS Solar Cell etc.) for various applications.

In 1954, the first solar panel has been invented in Bell Labs. Those first silicon solar cells were about 6% efficient during solar energy conversion to electricity. After 1960, solar cells with up to 14% efficiency were found. Few researchers working at the University of New South Wills, Australia, got the result of 20% efficiency around 1985.

Recently, the Sharp and National Renewable Energy Laboratory (NREL) experimenters achieved 44.4% efficiency using silicon-based material. The average efficiency of Silicon-based Solar cells lies between 16-18%, where the maximum on-field efficiency is due to the advancement with PERC (Passivated Emitter Rear Contact) technology in the discrete back surface mode for the collection of holes. This technology provides efficiency up to 22% in Polycrystalline modules [6]. Due to silicon's low price and good stability and efficiency, it is used widely for photovoltaic applications. The photovoltaic modules are made up of the combination of various cells in series/parallel to each other to provide the desired power. Furthermore, these modules are connected at string level with other modules to combine their power and fulfill the overall energy requirement [7].

The solar generation is directly proportional to the amount of light falling on a solar cell/panel. The increase/decrease in light intensity will directly affect the current generation of the particular solar cell, while the voltage variations due to change in intensity would be minimum. Generally, the variation in light intensity is formed due to partial shading, environmental changes (cloud variations), dust, and any objects that are fallen on a particular solar cell.

As mentioned in [9], three types of combinational analysis are done, which are Series-Parallel (SP), Bridge Linked (BL) and, Total Cross Tied (TCT) methods of connecting the solar cells while making the modules. The conclusion made as per this experiment was – the maximum output could be recorded using the total Cross Tied (TCT) configuration. Also, the MPPT (Maximum Power Point Tracking) technique is used to track the system performance and regulate the voltages accordingly [8, 9].

As mentioned in [10], the partial shading is happening due to soil shading on solar panels. These soil shading is of two types namely – hard soil and soft soil shading. Soft soil shading occurs due to air pollution, and hard soil shading occurs due to dust particles settled down on the panels as it blocks sunlight to enter within it. As per the survey conducted in Boston, the 1% solar irradiation loss occurred due to dust accumulated on the solar panel with a tilt angle of 30°, while the highest degradation has been found up to 4.7%. To overcome this issue, they have proposed manual cleaning and mobile cleaning method of solar panels. The manual cleaning deals with using a ladder and a scrub to clean the solar panel manually, while the mobile cleaning deals with the cleaning of solar photovoltaic modules with help of water sprinklers.

Presently, bypass diodes are placed in solar PV panels, but the drawbacks of using bypass diodes are forward voltage drops and hotspots. To overcome it, electromagnetic relays are used. As per the datasheet of the silicon diode, it offers a 0.6 V forward voltage drop, so if the solar panel is rated for 20 V and 40 W; then it will generate about 2A of current, so power loss in the diode will be 2 W. By the use of bypass diode for an average condition loss of 550 mW power occurs at 0.62 A. According to experimental results, a shallow loss of control of 150 mW happened under the same shading conditions by using electromagnetic relays [11, 12].

It was also observed that a more significant number of bypass diodes are used to get higher efficiency and maximum power output during shading conditions. Also, the performance depends on the shape of the shading pattern. It was also concluded that once a bypass diode is activated, it acts as a different route for the current [13]. The variation in irradiation on solar panels reduces the solar panel's output and thus results in mismatch losses. To overcome this, reconfiguration by the Total Cross Tied (TCT) method is used, and it is divided into different stages. In the first stage, the PV plant is switched between different arrays, and dispersion is done in the second stage; distribution is done by switching between columns. By this algorithm, the quantity of switches and sensors is also reduced. The conclusion was made by studying four cases that two stages generate more power in partial shading conditions than the Total Cross Tied (TCT) method [14].

It has been found that shading creates a hotspot problem not only in the shaded cells but in the unshaded cells also. By shading half of a single cell, the power curtailment was 25.71%. Simultaneously, when 50% of the panel was shaded, power reduction was found to be 70.27%. For the same shading condition in series and parallel configurations, there was a reduction of 16.54% and 6.03%, respectively [15]. The simulation was made in the software named MATLAB/SIMULINK, and it was verified with accurate measured datasets [16].

The experimental result found that a particular PV System provides 2373 kWh power with no shading (maximum output condition) mode. It was recorded that the system produces 2055 kWh power at 2.1% shading. It was observed that the reduction in the output depends upon the shaded area and the position at which the radiation reaches within the solar cell. The below-mentioned Table 13.1 consists of the experimental result and illustrates the data such as shadow percentage, current, voltage, and power production while keeping constant irradiance value in real time domain [17].

The experiment is all about the classification of shades, in which the shadows are divided into two parts, namely partial shading and whole

Table 13.1 Comparative power generation analysis with variable shadow percentage [17].

Irradiance (%)	Shadow (%)	Current (Amp)	Voltage (V)	Power (Watt)
100	0	3.95	364	1437
100	20	3.48	374	1301
100	50	2.6	370	962
100	100	0.00012	183	0.02196

shading. Further partial shading is also classified as uniform shading and non-uniform shading as shown in Figure 13.1.

The Subjective shading is classified into two categories static shading and dynamic shading. Using different non-uniform shadings, the experiment was performed to record the efficiency and area of coverage. The bird model had an area of 4% coverage, and the output efficiency found was 3.21%. Similarly, a building model had a coverage area of 19.82%, and the efficiency observed was 2.46%. The conclusion was found that an effective method to reduce shading-related consequences was of bypass diode method [18].

As mentioned in Bulanyi, P. and Zhang, R. [19], different shading scenarios and classification and the use of bypass diode functionality are investigated in detail. The experiment recommends the reduction in shading losses through installation solution. It was observed that when each cell was partially shaded by 25%, there was a drop in current, but the voltage remained unchanged. When there is shading, the adjacent cell voltages forward bias its bypass diode, and such a condition is known as the active state. Here the impedance becomes lower and current flows through it. Hence overheating hotspot problem is resolved. Site inspection is done using SolmetricSun Eye 210, a method to compute shading accurately. Landscape orientation is also a method to reduce module losses.

Two techniques are used to reduce power losses during partial shading, namely active and passive techniques. Where the passive approach is all about with primary bypass diode. The Distributed Maximum Power Point Tracker (MPPT) technique, multilevel inverters, and reconfiguration strategies are the three factors classified by the active methods. Active techniques are more convenient for reducing losses during partial shading conditions [20]. Also, the image processing like median filtering and morphological functions (using structuring elements) to have required outputs are used in some modules to prevent the shading effect [21].

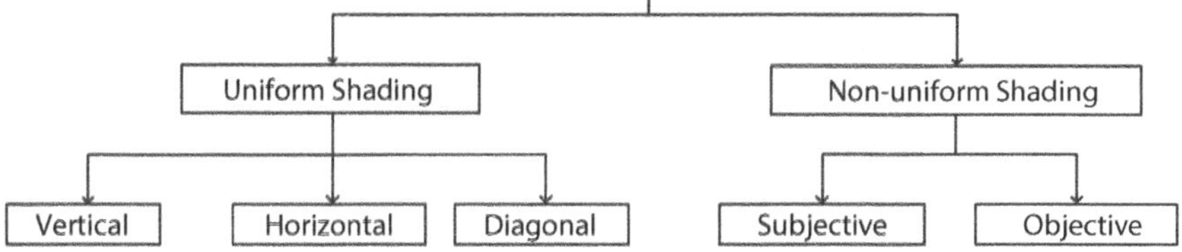

Figure 13.1 Types of partial shading occurring on solar photovoltaic system.

This manuscript is divided into four parts, Section 13.1 is the introductory part about shading in photovoltaic system, Section 13.2 is the proposed methodology and some mathematical formulation that is getting affected by shading and hotspots. The Section 13.3 is the results and discussion of our proposed methodology while the Conclusions, Acknowledgement, COI and References are mentioned in Section 13.4.

13.2 Proposed Methodology

The solar cells generates electricity when light falls on a cell. Illuminated solar cell's IV curve is given by Shockley's equation [22]:

$$I = I_0\left(exp\frac{eV}{kT} - 1\right) \tag{13.1}$$

$$I_0 = qA\left(\frac{n_i^2}{N_A}\frac{L_e}{\tau_e}\xi_e + \frac{n_i^2}{N_D}\frac{L_h}{\tau_h}\xi_h\right) \tag{13.2}$$

$$\xi_e = \frac{s_e cos\,h(W_A / L_e) + (L_e / \tau_e) sinh(W_A / L_e)}{(L_e / \tau_e) cosh(W_A / L_e) + s_e sinh(W_A / L_e)} \tag{13.3}$$

$$\xi_h = \frac{s_h cos\,h(W_C / L_h) + (L_h / \tau_h) sinh(W_C / L_h)}{(L_h / \tau_h) cosh(W_C / L_h) + s_h sinh(W_C / L_h)} \tag{13.4}$$

Where, I = net current flowing through diode, I_0 = dark saturation current (diode leakage current density in absence of light), V = applied voltage across diode terminal, e = absolute value of electron charge ($1.6 * 10^{-19}$ C), k = Boltzmann Constant ($1.38 * 10^{-23}$ $m2\ kgS^{-2}\ k^{-1}$), T = absolute temperature of diode junction, W_A = diode anode thickness, W_C = diode cathode thickness, L_e = electron diffusion length, L_h = hole diffusion length, n_i = intrinsic carrier concentration, N_A = acceptor concentration in electron, A = device area, τ_e = electron lifetime, τ_h = hole lifetime, s_e = recombination velocity of electron surface, s_h = recombination velocity of hole surface, q = electron charge, and N_D = net donor concentration in the cathode.

The intrinsic career concentration (n_i) and its temperature dependence expression mathematically can be written as [23, 24] –

$$n_i = \alpha T^{3/2} exp\left(\frac{-Eg}{2k_B T}\right) \tag{13.5}$$

$$n_i = \sqrt{N_A N_D} exp\left(\frac{-|e|V_{bi}}{2k_B T}\right) \tag{13.6}$$

The absorption coefficient is a constant depending on the effective mass present in conduction and valence bands of electrons and holes. At the same time, the bandgap has some dependency on variation in temperature. These temperature variations can be caused due to the mismatch/presence of the hotspot region in the solar photovoltaic module. Furthermore, the absorption is dependent on various parameters [25]:

$$\alpha(T) = \sum_{\substack{i=phonon\,1,2 \\ j=band\ gap\,1,2}} A_j C_i \frac{[hv - E_{g,j}(T) + E_{ph,i}]^2}{exp(E_{ph,i}/kT) - 1} + \frac{[hv - E_{g,j}(T) - E_{ph,i}]^2}{1 - exp(E_{ph,i}/kT)} + \frac{[A_d(hv - E_{g,d}]^{3/2}}{hv} \tag{13.7}$$

Where, the i and j are phonons mentioned as 1 and 2, while the indirect bandgaps 1 and 2 are having a values of $E_{g,1}$ = 1.16 ev and $E_{g,2}$ = 2.25 ev. Also, the direct bandgap 3 is having the value of $E_{g,d}$ = 3.2 ev at 0 K temperature. Also, the $E_{ph,1}/K = 212$ K and $E_{ph,2}/K = 670$ K are the phonons with C_1 = 5.5 and C_2 = 4.0. In the above equation, the first term is explaining about the absorption of phonons, while the second and third terms explain the emission of phonons during the absorption of light and absorption of a direct (forbidden) transition, respectively.

$$E_g(t) = E_g(0) - \frac{\beta T^2}{T + \gamma} \tag{13.8}$$

By using above equation, the temperature dependent bandgap can be calculated, Where the $\beta = 7.021 \times 10^{-3}$ eV/K and $\gamma = 1108$ K.

The equation (13.1) applies to the diode working in the ideal case. During illumination on a surface of the cell will cause light generated current, and under that condition, the conventional diode equation can be modified with [26] -

$$I = I_L - I_0\left(exp\frac{eV}{nkT} - 1\right) \tag{13.9}$$

Where, I_L = light generated current and n = ideality factor. But this equation can further be modified by considering the series and shunt resistance of solar cell [27] –

$$I = I_L - I_0\left[exp\left(\frac{(V + IR_s)}{V_t}\right) - 1\right] - \frac{V + IR_s}{R_{sh}} \tag{13.10}$$

Where, $V_t = aKT_c/q$ (V_t is the thermal voltage), I = cell output current, R_s = series resistance, R_{sh} is the shunt resistance and V is the cell output voltage. As mentioned in Volker Q. [28], the shading effect on solar cells heavily affects the short circuit current while the rate of change in open circuit voltage v_{OC} is less. This phenomena happens due to direct dependency of light generated current (I_L) on the short circuit current (I_{SC}) of a solar cell. During full sun condition, the light generated current will be equivalent to short circuit current. Mathematically this can be represented as: $I_L \approx I_{SC}$. While variation in irradiance can directly affect the I_L that will further affect the value of I_{SC}. While I_L is dependent in logarithmic function with v_{OC}, which can be written as:

$$v_{OC} = \frac{nkT}{q} ln\left(\frac{I_L}{I_0} + 1\right) \tag{13.11}$$

To overcome the issue of partial shading, we have divided the 12 cell-based modules into a group of three, named G_1, G_2, and G_3. All three groups contain 12 solar cells connected in series, which are equally divided into a group of 4 cells, individually forming three groups. The circuit diagram of the entire panel would look like.

Figure 13.2(a) is the schematic representation of a commercially available photovoltaic module with 72 series-connected cells in a pair of 6 containing 12 cells each. Figure 13.2(b) is the modified version of the commercially available photovoltaic module mentioned in Figure 13.2(a), including 72 cells that are connected in series containing six pairs of 12 cells. Furthermore, each pair comprising three groups named G_1, G_2 and G_3 (mentioned in Figure 13.2(c)) individually contains four series-connected solar cells.

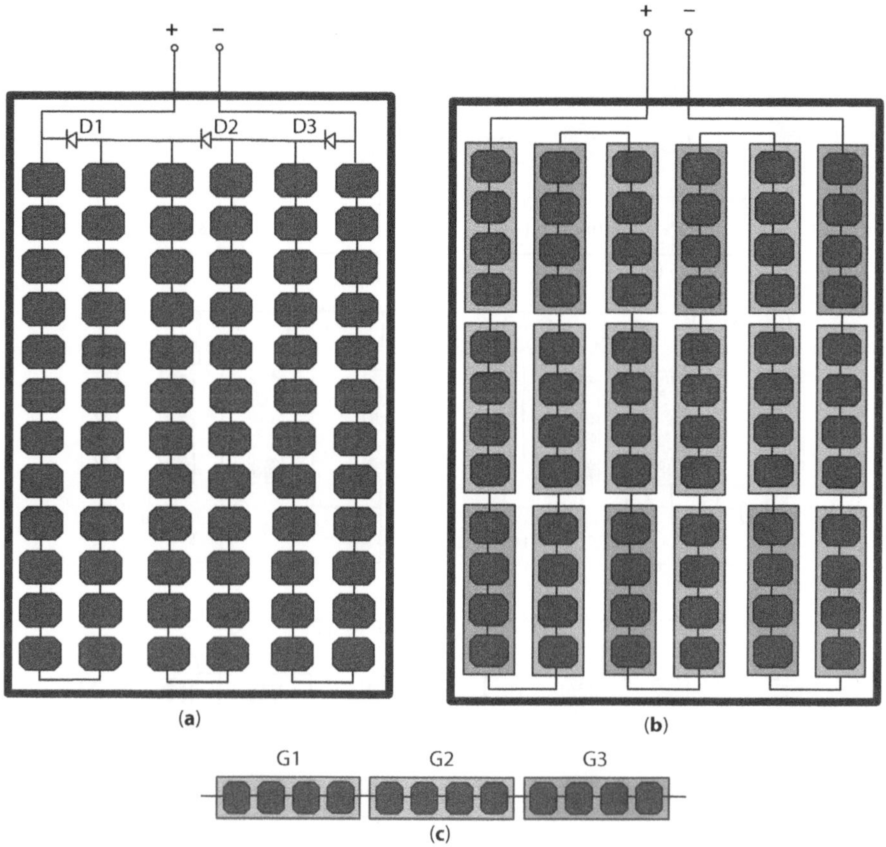

Figure 13.2 Illustration of solar modules and its modifications in groups. (a) 72 series connected cells. (b) modified version of cells with controlled architecture. (c) individual connections of groups.

Figure 13.2(c) has 12 bypass switches for each pair, unlike the three diodes available in commercial photovoltaic modules. The individual working operation of each switches are further dependent on the shadow level by which multiple groups are getting affected. Out of six available pairs, the study of one pair containing three groups is working for the cases mentioned below.

Figure 13.3 (a-h) shows the connection diagram for 8 cases, as mentioned in Table 13.2. The modified solar modules mentioned in Figure 13.2(b) containing six pairs with three groups in each named G_1, G_2 and G_3 are represented. The one pair representation has eight possible states mentioned in Table 13.2. Also, its connection for the eligible cases is drawn in Figure 13.3 (a-h). Figure 13.3(a) is the connection diagram of the case1 mentioned in Table 13.2, the Figure 13.3(b) is the connection diagram

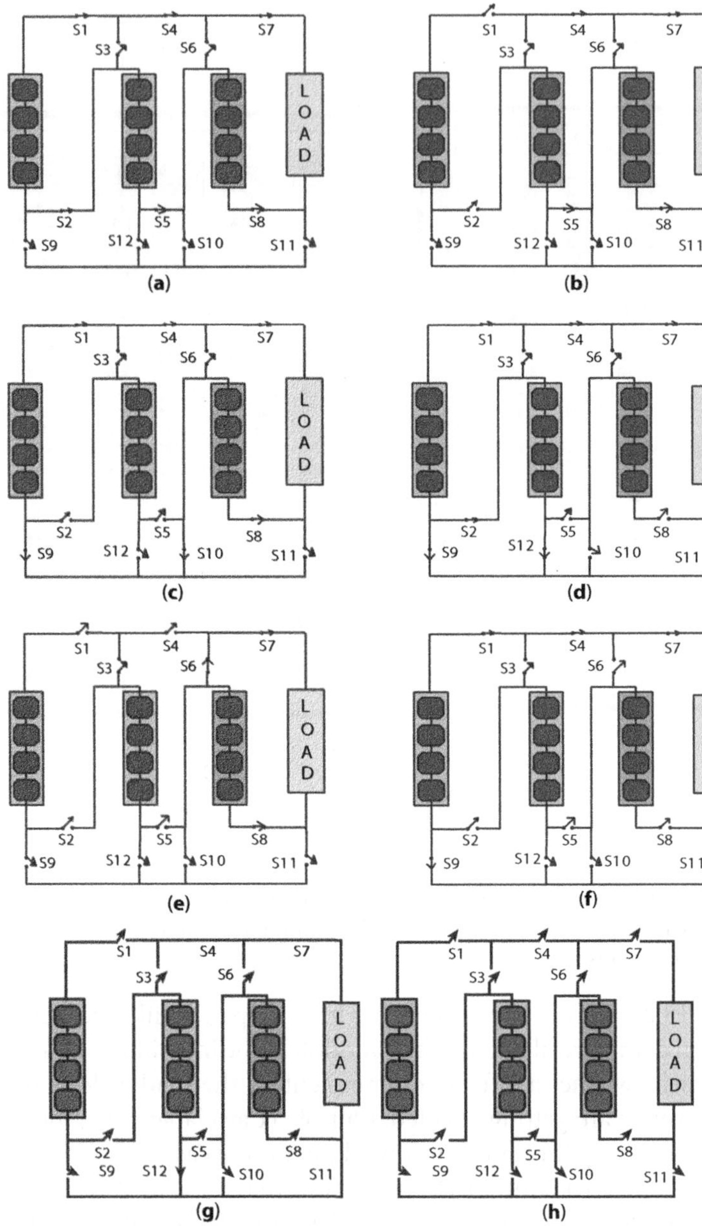

Figure 13.3 (a) Connection diagram of case 1 mentioned in Table 13.2. (b) Connection diagram of case 2 mentioned in Table 13.2. (c) Connection diagram of case 3 mentioned in Table 13.2. (d) Connection diagram of case 4 mentioned in Table 13.2. (e) Connection diagram of case 5 mentioned in Table 13.2. (f) Connection diagram of case 6 mentioned in Table 13.2. (g) Connection diagram of case 7 mentioned in Table 13.2. (h) Connection diagram of case 8 mentioned in Table 13.2.

Table 13.2 Operation of switches for various cases.

Case no.	Description	On switches	Off switches
1	G_1,G_2,G_3 will be in healthy mode	S_1,S_2,S_4,S_5, S_7,S_8,	S_3,S_6,S_9,S_{10},S_{11},S_{12}
2	G_1 is unhealthy while G_2 and G_3 are healthy	S_4,S_5,S_7,S_8	S_1,S_2,S_3,S_6,S_9, S_{10},S_{11},S_{12}
3	G_2 is unhealthy while G_1 and G_3 are healthy	S_1,S_4,S_7,S_8, S_9,S_{10}	S_2,S_3,S_5,S_6, S_{11},S_{12}
4	G_3 is unhealthy while G_1 and G_2 are healthy	S_1,S_2,S_4,S_7, S_9,S_{11},S_{12}	S_3,S_5,S_6, S_8,S_{10}
5	G_1 and G_2 are unhealthy while G_3 is healthy	S_6,S_7,S_8	S_1,S_2,S_3,S_4, S_5,S_9,S_{10},S_{11},S_{12}
6	G_2 and G_3 are unhealthy while G_1 is healthy	S_1,S_4,S_7,S_9,S_{11}	S_2,S_3,S_5,S_6, S_8,S_{10},S_{12}
7	G_1 and G_3 are unhealthy while G_2 is healthy	S_4,S_7,S_{11}, S_{12}	S_1,S_2,S_3,S_5, S_6,S_8,S_9,S_{10}
8	G_1,G_2 and G_3 are unhealthy	-----------	S_1,S_2,S_3,S_4,S_5,S_6 S_7,S_8,S_9,S_{10},S_{11},S_{12}

of case2. Similarly, the remaining Figures 13.3(c, d, e, f, g and h) are the connection diagrams of the case (3, 4, 5, 6, 7 and 8), respectively mentioned in Table 13.2.

One pair containing three groups needs 12 power electronics (controlled switches) to achieve the maximum energy generation at the module level. Here, in this experiment, we have used MOSFET (Metal Oxide Semiconductor Field Effect Transistor) switches to provide a bypass at the individual group level during the case of shading. These switching effects are performed due to the gate pulse given by the microcontroller by sensing and comparing the current values I_1, I_2 and I_3 of individual groups G_1, G_2 and G_3.

If each cell is getting the radiation of 100%, then the value of current I_1, I_2 and I_3 flowing through all the three groups G_1, G_2 and G_3 will be equal, and the resultant will lead to the healthy condition of all groups. This will be the type of Case1 mentioned in Table 13.2, and it will cause the triggering of switches S_3, S_6, S_9, S_{10}, S_{11}, and S_{12}, while the switches S_1, S_2, S_4, S_5, S_7, and S_8 will be deactivated. In this condition, all the energy through each group will be utilized without bypassing any particular group.

If the first group is shaded, then the current I_1 flowing through G_1 will be less than the current I_2 and I_3 flowing through G_2 and G_3. During this, switches S_1, S_2, S_6, S_9, S_{10}, S_{11}, and S_{12} will be triggered, while S_3, S_4, S_5, S_7, and S_8 will be deactivated. This category will be the type of case2, where the energy of G_2 and G_3 will be utilized by simply bypassing the energy of G_1.

If the second group is shaded, then the current I_2 flowing through G_2 will be less than the current I_1 and I_3 flowing through G_1 and G_3. During this, switches S_2, S_3, S_5, S_6, S_{11} and S_{12} will be triggered, while S_1, S_4, S_7, S_8, S_9 and S_{10} will be deactivated. This category will be the type of case3, where the energy of G_1 and G_3 will be utilized by simply bypassing the energy of G_2.

If third group is shaded, then the current I_3 flowing through G_3 will be less than the current I_1 and I_2 flowing through G_1 and G_2. During this, switches S_3, S_5, S_6, S_8, S_9 and S_{10} will be triggered, while S_1, S_2, S_4, S_7, S_{11} and S_{12} will be deactivated. This category will be the type of case4 where the energy of G_1 and G_2 will be utilized by simply bypassing the energy of G_3.

If first and second group is shaded, then the current I_1 and I_2 flowing through G_1 and G_2 will be less than the current I_3 flowing through G_3. During this, switches S_1, S_2, S_3, S_4, S_5, S_9, S_{10}, S_{11}, and S_{12} will be triggered while S_6, S_7 and S_8 will be deactivated. This category will be the type of case5 where the energy of G_3 will be utilized by simply bypassing the energy of G_1 and G_2.

If second and third group is shaded, then the current I_2 and I_3 flowing through G_2 and G_3 will be less than the current I_1 flowing through G_1. During this, switches S_2, S_3, S_5, S_6, S_7, S_8, S_{10}, S_{11} and S_{12} will be triggered, while S_1, S_4, and S_9 will be deactivated. This category will be the type of case6 where the energy of G1 will be utilized by simply bypassing the energy of G_2 and G_3.

If first and third group is shaded, then the current I_1 and I_3 flowing through G_1 and G_3 will be less than the current I_2 flowing through G_2. During this, switches S_1, S_2, S_5, S_6, S_8, S_9 and S_{10} will be triggered, while S_3, S_4, S_7, S_{11} and S_{12} will be deactivated. This category will be the type of case7, where the energy of G2 will be utilized by simply bypassing the energy of G_1 and G_3.

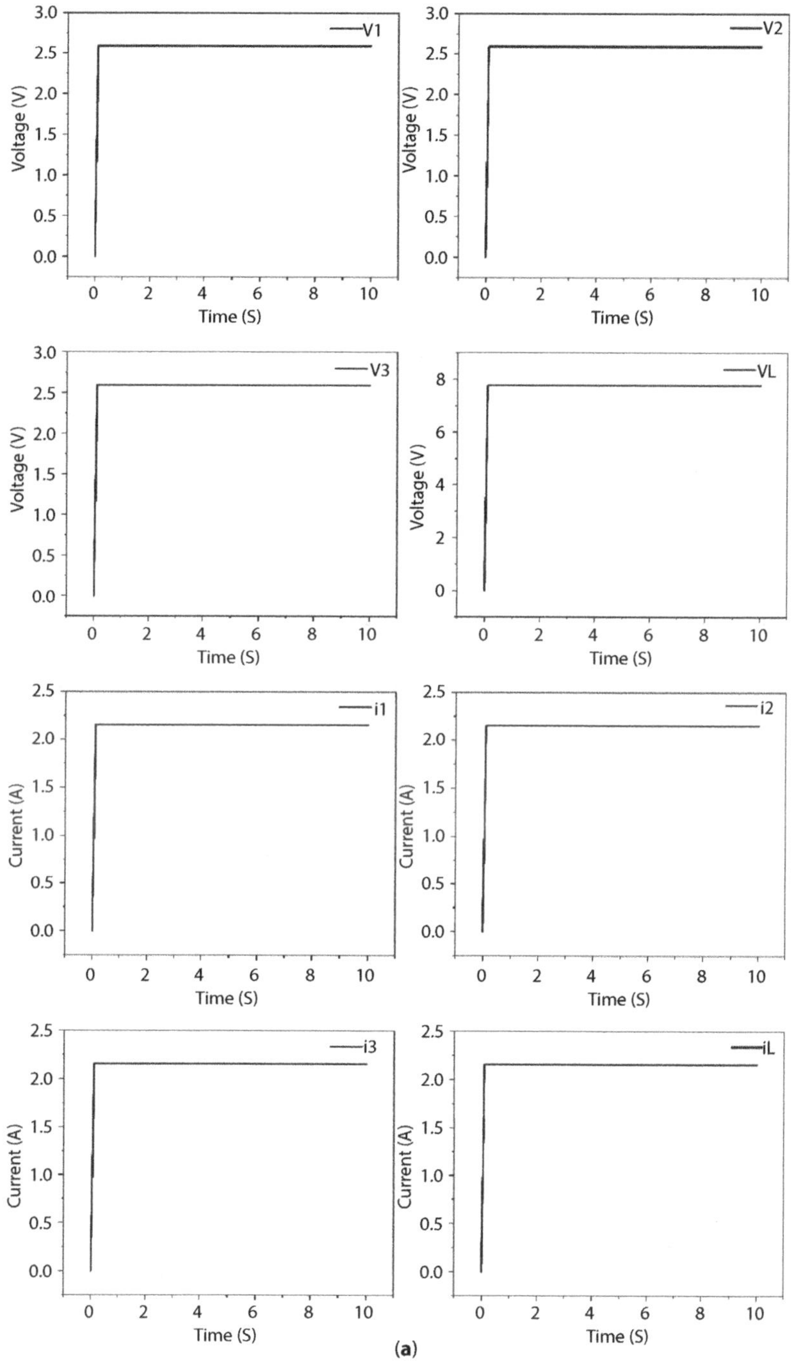

Figure 13.4 (a) Experimental result of case 1 mentioned in Table 13.2. *(Continued)*

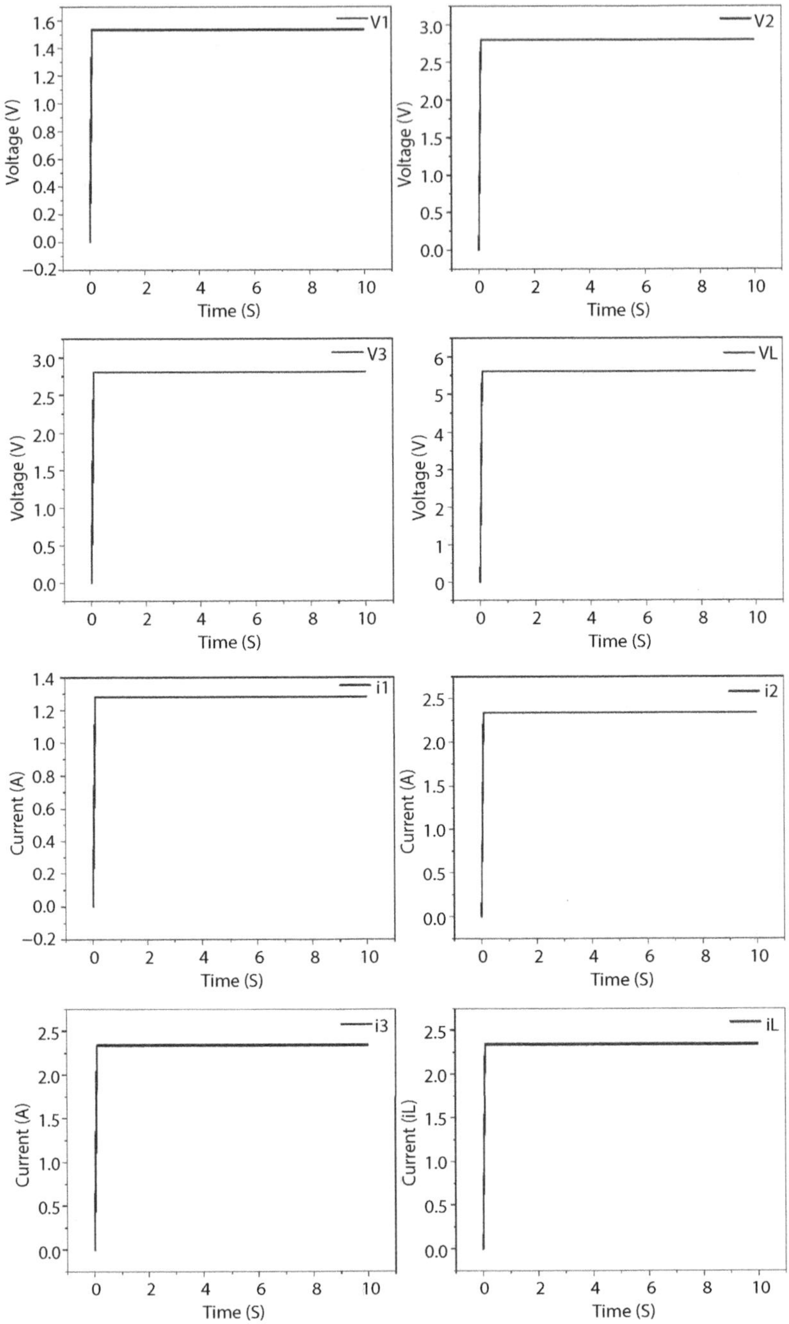

(**b**)

Figure 13.4 (Continued) (b) Experimental result of case 2 mentioned in Table 13.2.

(Continued)

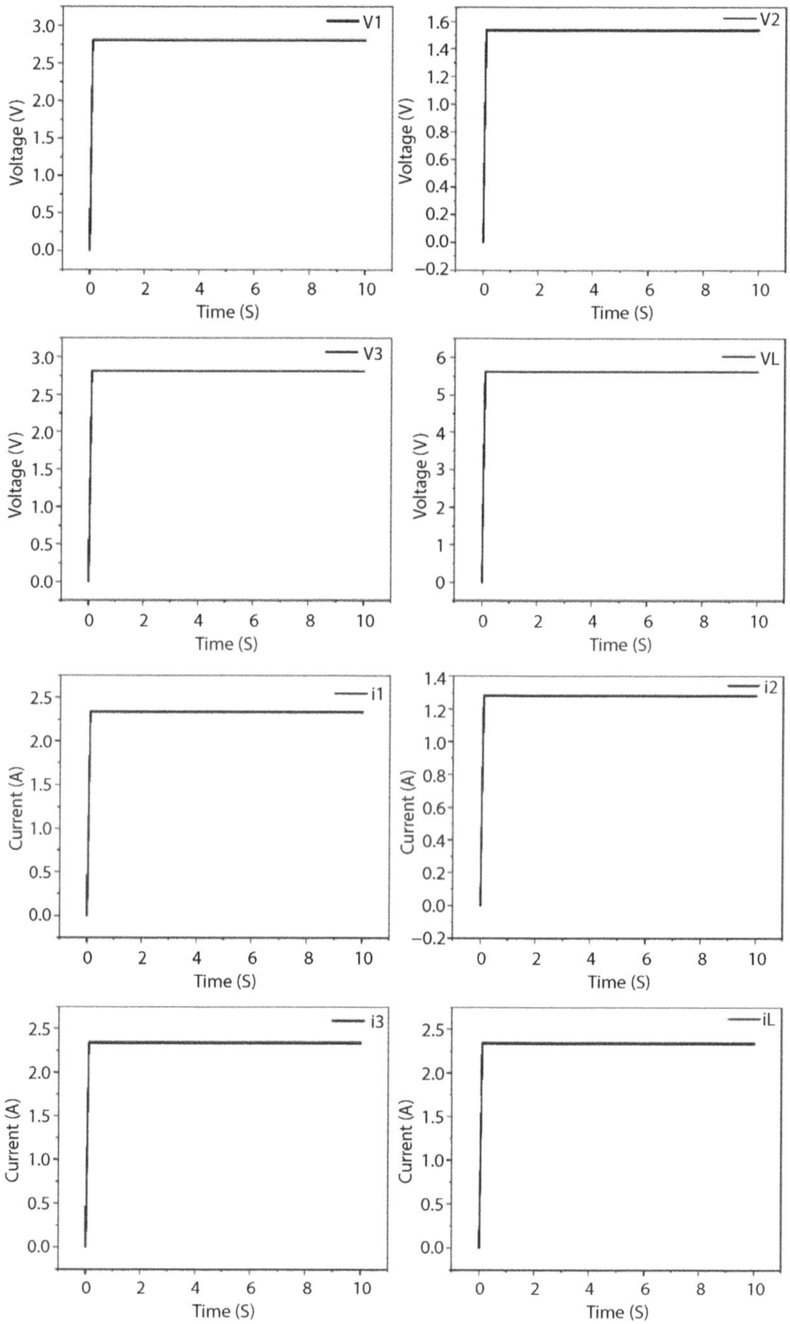

(c)

Figure 13.4 (Continued) (c) Experimental result of case 3 mentioned in Table 13.2.

(Continued)

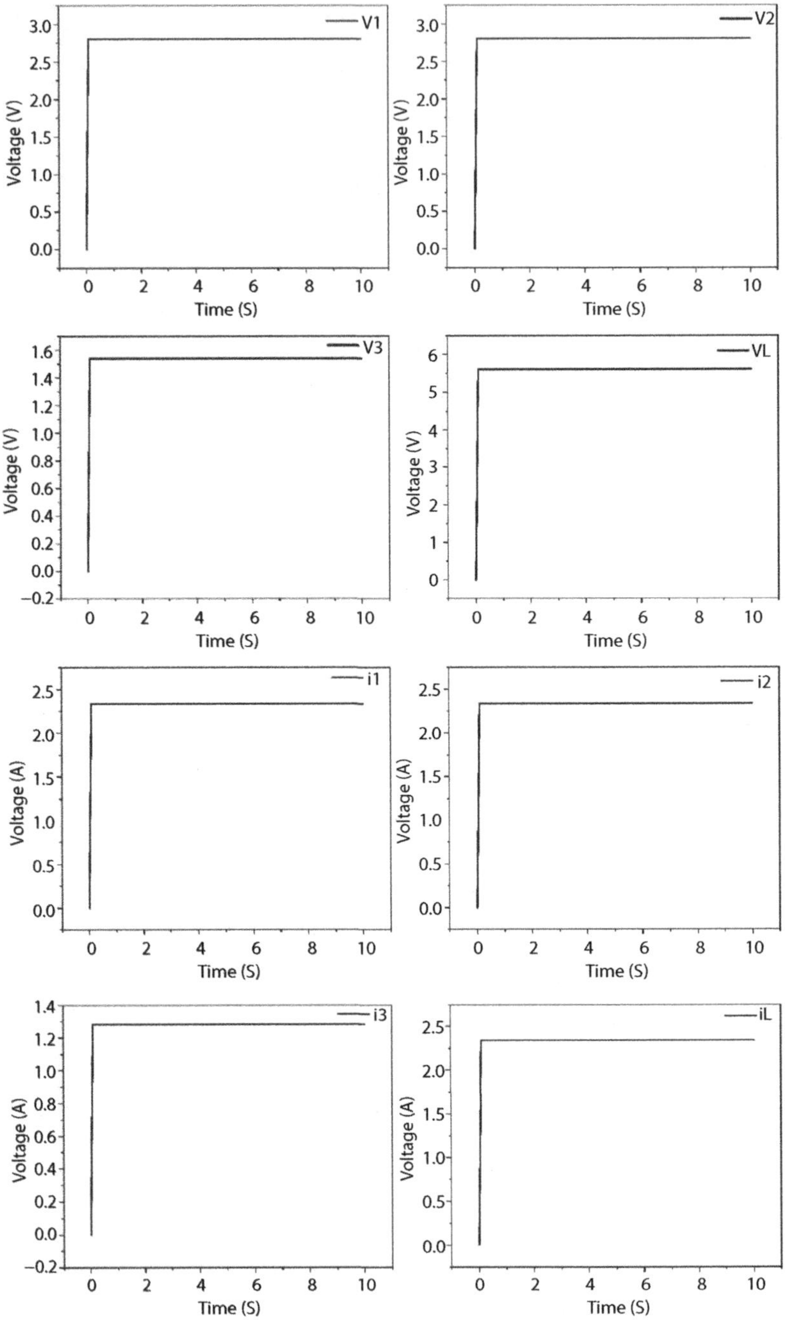

(d)

Figure 13.4 (Continued) (d) Experimental result of case 4 mentioned in Table 13.2.

(Continued)

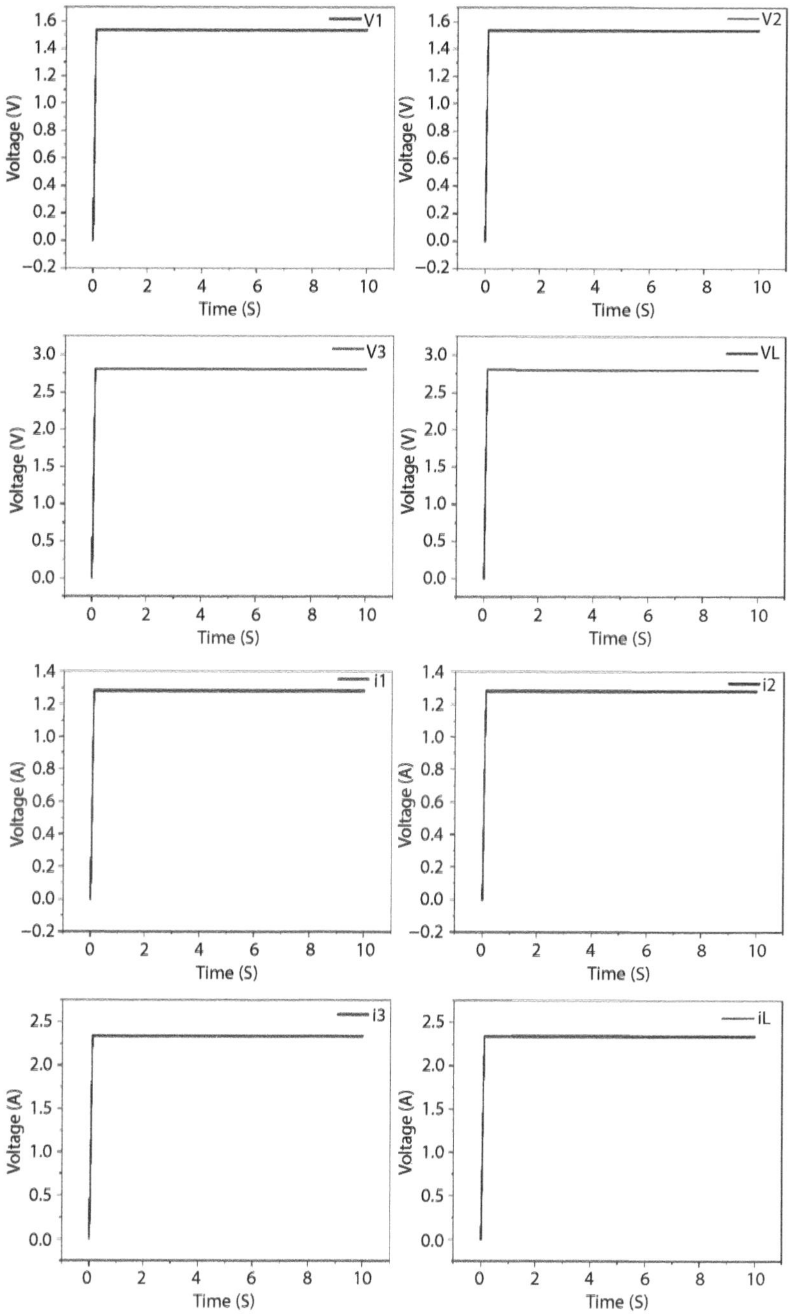

Figure 13.4 (Continued) (e) Experimental result of case 5 mentioned in Table 13.2.

(Continued)

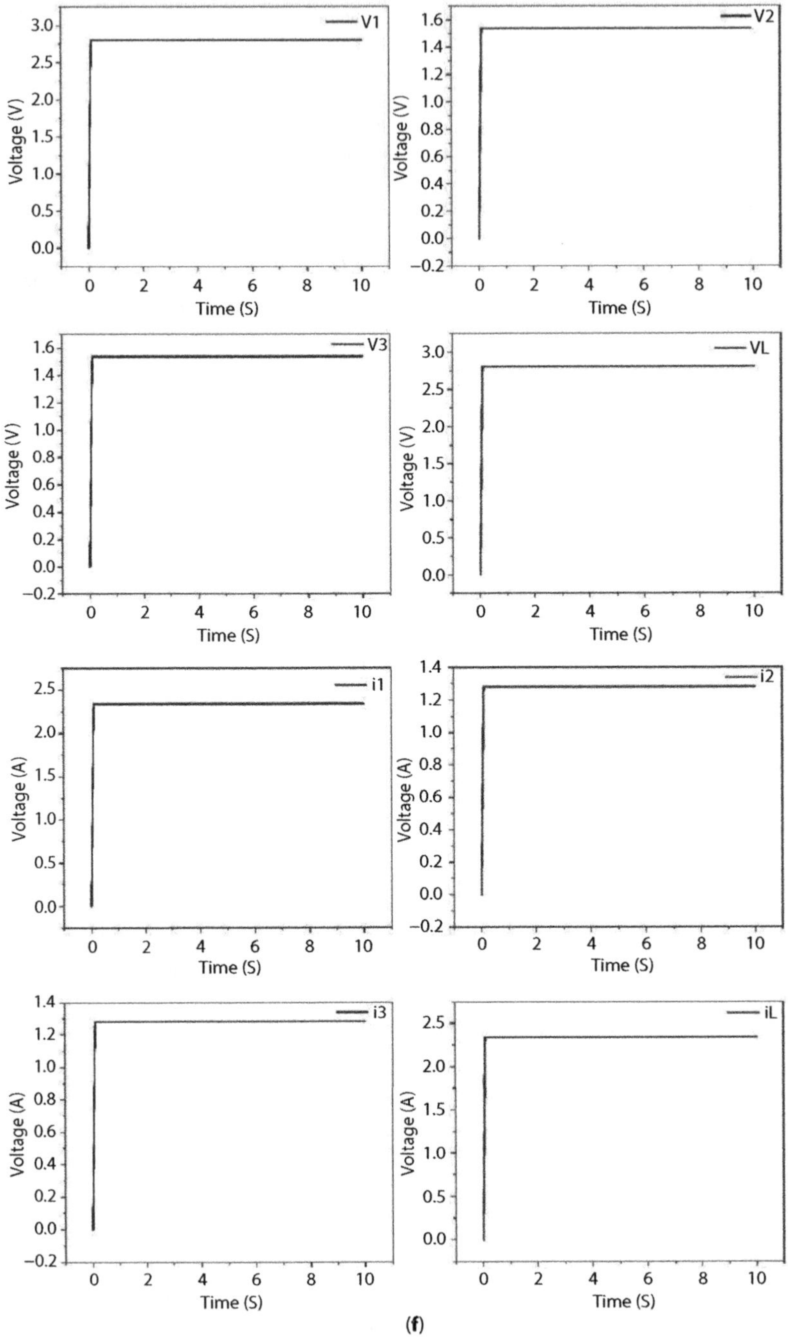

Figure 13.4 (Continued) (f) Experimental result of case 6 mentioned in Table 13.2.

(Continued)

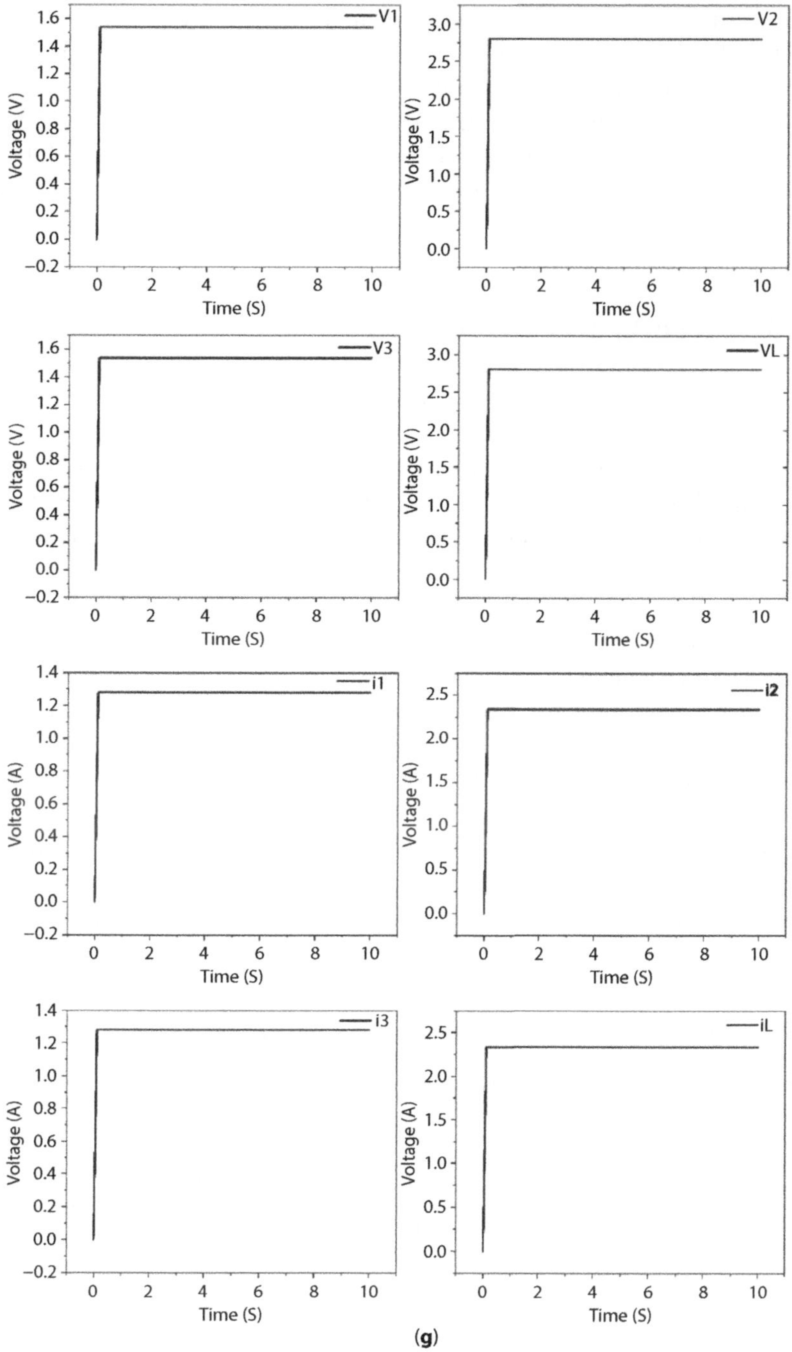

Figure 13.4 (Continued) (g) Experimental result of case 7 mentioned in Table 13.2.

If all the groups G_1, G_2 and G_3 is shaded, then the current I_1, I_2 and I_3 flowing through these groups will be having a very less current closed to minimum threshold current limit i.e., < 0.9% from the current flowing in healthy condition of operation. During this, all the switches S_1, S_2, S_3, S_4, S_5, S_6, S_7, S_8, S_9, S_{10}, S_{11} and S_{12} will be deactivated. This category will be the type of case8, where all three groups G_1, G_2 and G_3 will be simply bypassed without considering any energy.

13.3 Results and Discussion

The graphical representation of system response under variable irradiance data is illustrated below.

As mentioned above, Figure 13.4 represents the experimental results when the three groups G_1, G_2 and G_3 are simulated for one complete pair with 12 series connected cells. These series-connected cells are further connected with five more pairs containing 12 cells each, and resultant causes 72 solar cells connected in series with each other. The simulated cells had the following ratings.

These solar cells were simulated in SIMULINK library of MATLAB – R2015b Environment for two different types of radiation such as I_{R1} = 1000 W/m^2 during healthy/unshaded conditions and I_{R2} = 175 W m^2 during unhealthy/shaded conditions, respectively. And following values are observed.

Conventionally, 24 cells are connected in series with a pair of three and separated by three blocking diodes. If all the cells across the panel will have constant irradiance, then it will generate constant current while the

Table 13.3 Internal parameters of simulated solar cells.

Sr. no.	Parameters	Ratings
1	Short Circuit Current (I_{sc})	7.34 A
2	Open Circuit Voltage (V_{oc})	0.6 V
3	Irradiance for Measurement (I_{ro})	1000 W/m^2
4	Quality Factor (N)	1.5
5	Series Resistance (Ω)	0 Ω
6	Energy Gap (E_g)	1.11 ev

Table 13.4 Experimental values of three different groups named as: G_1, G_2 and G_3.

Case no.	V_1 (V)	V_2 (V)	V_3 (V)	I_1 (A)	I_2 (A)	I_3 (A)	V_L (V)	$_{iL}$ (A)	I_{R1} (W/m^2	I_{R2} (W/m^2	I_{R3} (W/m^2)
1	2.59	2.59	2.59	2.15	2.15	2.15	7.77	2.15	1000	1000	1000
2	1.53	2.80	2.80	1.28	2.33	2.33	5.61	2.33	175	1000	1000
3	2.80	1.53	2.80	2.33	1.28	2.33	5.61	2.33	1000	175	1000
4	2.80	2.80	1.53	2.33	2.33	1.28	5.61	2.33	1000	1000	175
5	1.53	1.53	2.80	1.28	1.28	2.33	2.80	2.33	175	175	1000
6	2.80	1.53	1.53	2.33	1.28	1.28	2.80	2.33	1000	175	175
7	1.53	2.80	1.53	1.28	2.33	1.28	2.80	2.33	175	1000	175
8	1.53	1.53	1.53	1.28	1.28	1.28	00	00	175	175	175

potential difference of each cells will be simply added up. But if anyone cell is having a shadow, it will be bypassed through blocking diode. In that case, energy generation with 24 cells will be simply bypassed if any cell is shaded.

To tackle this issue, the group architecture and its performance mentioned in Table 13.4 will cause more generation during the shading effect. In the table, the group G_1, G_2 and G_3 voltages V_1, V_2 and V_3, along with currents I_1, I_2 and I_3 are observed during the simulation for three irradiance I_{R1}, I_{R2} and I_{R3} values falling on each group. The effects of these fluctuations are clearly observable at the last point in the form of voltage *(V_L)* and Current *(i_L)*.

If all the groups are connected in conventional mode, the value of power at the last point of each 24 cells will be –

$$P_L = V_L \times I_L \tag{13.12}$$

$$V_L = \sum_{i=1}^{N} V_i \tag{13.13}$$

As per Table 13.4, if one cell at full radiation tends to produce a voltage of 2.59 *V* and a current of 2.15 *A*, and if they are connected in series with *N = 24* cells, then according to equation 13, its voltages will be added up while the current at last point will be minimum current flowing through any one of the cell. If all the cells are in healthy condition and generating constant energy then the voltage across last point will be $V_L = 62.16\ V$ while the current flowing in last point will be $i_L = 2.15$ A. And according to equation 12, it will generate a power P_L *of 133.64 W*. If cells are connected conventionally, then 24 cells will have a power of 133.64W, and similarly the 72 cells will be having a power at the last point of module $P_L = 400.92W$ with $V_L = 186.48V$ and $i_L = 2.15A$. But if any cell from the group of 24 series connected architecture is has a shadow effect. Due to the inclusion of bypass diode, 24 cells will be bypassed and at the last point of module the P_L will be available as *267.28W* with $V_L = 124.32V$ and $i_L = 2.15A$. Similarly, if two groups containing 24 cells will have shadow effect, the 48 series connected cells will be bypassed and the last point of module the P_L will be available as *133.64W* with $V_L = 62.16V$ and $i_L = 2.15A$.

Considering the above issue, the 72 cell connections are modified in a set of 6 pairs containing three groups, each with four series-connected cells. These groups are connected in series with each other and form a pair, and the pairs are also connected in series with remaining pairs. If all the cells are in healthy condition and generating constant energy, then the voltage across the last point of the module will be $V_L = 186.48$ V, while the current flowing in the last point will be $i_L = 2.15$ A. And according to equation 12, it will generate a power P_L *of 400.92 W*. But if any cell of any one group containing four cells will be affected by shadow, the voltage across last point of module will be $V_L = 176.12$ V, while the current flowing in the last point will be $i_L = 2.15$ A. And according to equation 12, it will generate a power P_L *of 378.65 W.* In this case, the power generation is boosted up because only four cells are bypassed compared to the previous architecture where 24 cells were bypassed. At this moment, **27.78%** more energy is utilized.

Similarly, if two groups are getting shaded of pairs one and three compared with two groups of 48 cells (conventional architecture); the 48 cells will be bypassed and will have the power (P_L) at the last point of the module as *133.64W* with $V_L = 62.16V$ and $i_L = 2.15A$. While in the case of two groups from pair one and pair three, only the eight cells will be bypassed and will have the power (P_L) at the last point of the module as *356.384W* with $V_L = 165.76V$ and $i_L = 2.15A$. This modification will cause the utilization of **55.56%** more energy than conventional architecture with the presence of a blocking diode.

13.4 Conclusion

In this research, the conventional architecture of 72 cells containing module is modified with six pairs where individual pairs have three groups named G_1, G_2 and G_3. The experimental results show that the group-based arrangement is more efficient than the conventional based architecture of 72 series-connected modules with three blocking diodes. According to experimental results, the group-based architecture provides 27.78% more efficient energy generation than the traditional architecture of the module during one cell shading. While the two different location shading caused the increased energy generation by 55.56%. The experimental results show that, the generation efficiency is utilized more when the shading effect increases compared with the conventional architecture of 72 cells.

Acknowledgement

The authors would like to thank team MathWorks for MATLAB R-2015b Software for simulating the experimental setup, the ORIGIN and Microsoft VISIO Software for providing good plotting and graphical environment.

References

1. Lee, J.H. and Ramamurthi, K., *Fundamentals of Thermodynamics*, CRC Press, Boca Raton, Florida, 1998.
2. Pandey, S.V., Patel, J., Dhiman, H.S., Battery state-of-charge modeling for solar PV array using polynomial regression, in: *Artificial Intelligence for Renewable Energy Systems*, pp. 115–128, 2022.
3. Breeze, P., *Power Generation Technologies*, Newnes, Chennai, India, 2019.
4. Gross, R., Leach, M., Bauen, A., Progress in renewable energy. *Environ. Int.*, 29, 1, 105–122, 2003.
5. Suhagiya, J., Raval, D., Pandey, S.V., Patel, J., Gupta, A., Srivastava, A., Forecasting of electricity consumption for G20 members using various machine learning techniques, in: *Artificial Intelligence for Renewable Energy Systems*, pp. 211–227, 2022.
6. Pandey, S.V., Ganapuram, S., Singh, A.K., A review on advancements in photovoltaic cells, in: *Advances in Interdisciplinary Engineering*, pp. 227–236, Springer, Singapore, 2021.
7. Kazem, H.A., Chaichan, M.T., Alwaeli, A.H., Mani, K., Effect of shadows on the performance of solar photovoltaic, in: *Mediterranean Green Buildings & Renewable Energy*, pp. 379–385, Springer, Cham, 2017.
8. Farh, H.M. and Eltamaly, A.M., Maximum power extraction from the photovoltaic system under partial shading conditions, in: *Modern Maximum Power Point Tracking Techniques for Photovoltaic Energy Systems*, pp. 107–129, Springer, Cham, 2020.
9. Tubniyom, C., Chatthaworn, R., Suksri, A., Wongwuttanasatian, T., Minimization of losses in solar photovoltaic modules by reconfiguration under various patterns of partial shading. *Energies*, 12, 1, 24, 2019.
10. Maghami, M.R., Hizam, H., Gomes, C., Radzi, M.A., Rezadad, M., II, Hajighorbani, S., Power loss due to soiling on solar panel: A review. *Renew. Sust. Energ. Rev.*, 59, 1307–1316, 2016.
11. Gupta, A.K., Maity, T., Anandakumar, H., Chauhan, Y.K., An electromagnetic strategy to improve the performance of PV panel under partial shading. *Comput. Electr. Eng.*, 90, 106896, 2021.
12. Daliento, S., Di Napoli, F., Guerriero, P., d'Alessandro, V., A modified bypass circuit for improved hot spot reliability of solar panels subject to partial shading. *Sol. Energy*, 134, 211–218, 2016.
13. Pannebakker, B.B., de Waal, A.C., van Sark, W.G., Photovoltaics in the shade: One bypass diode per solar cell revisited. *Prog. Photovolt. Res. Appl.*, 25, 10, 836–849, 2017.
14. Krishna, G.S. and Moger, T., Reconfiguration strategies for reducing partial shading effects in photovoltaic arrays: State of the art. *Sol. Energy*, 182, 429–452, 2019.
15. Tripathi, A.K., Aruna, M., Murthy, C.S., Performance of a PV panel under different shading strengths. *Int. J. Ambient Energy*, 40, 3, 248–253, 2019.

16. Tabish, S. and Ashraf, I., Simulation of partial shading on solar photovoltaic modules with experimental verification. *Int. J. Ambient Energy*, 38, 2, 161–170, 2017.
17. Bernadette, D., Twizerimana, M., Bakundukize, A., Pierre, B.J., Theoneste, N., Analysis of shading effects in solar PV system. *Int. J. Sustain. Green Energy*, 10, 47–62, 2021.
18. Singh, N., Chauhan, Y.K., Singh, S.P., Srivastava, A.K., Shading issues in photo-voltaic modules and its performance assessment, in: *2018 International Conference on Power Energy, Environment and Intelligent Control (PEEIC)*, IEEE, pp. 280–287, April 2018.
19. Bulanyi, P. and Zhang, R., Shading analysis & improvement for distributed residential grid-connected photovoltaics systems, in: *The 52nd Annual Conference of the Australian Solar Council*, May 2014.
20. Krishna, G.S. and Moger, T., Reconfiguration strategies for reducing partial shading effects in photovoltaic arrays: State of the art. *Sol. Energy*, 182, 429–452, 2019.
21. Sarkar, G. and Bakade, K., Analysis of shading pattern of solar panels. *Int. J. Res. Eng. Technol.*, 3, 02, 594–599, 2014.
22. Martı, A., Balenzategui, J.L., Reyna, R.F., Photon recycling and Shockley's diode equation. *J. Appl. Phys.*, 82, 8, 4067–4075, 1997.
23. Garland, J.E., Crain, D.J., Roy, D., Impedance spectroscopy coupled with voltammetry for quantitative evaluation of temperature and voltage dependent parameters of a silicon solar cell. *Solar Energy*, 85, 11, 2912–2923, 2011.
24. LI, S.S., *Semiconductor Physical Electronics*, Plenum Press, New York, 1993.
25. Bücher, K., Bruns, J., Wagemann, H.G., Absorption coefficient of silicon: An assessment of measurements and the simulation of temperature variation. *J. Appl. Phys.*, 75, 2, 1127–1132, 1994.
26. Giebink, N.C., Wiederrecht, G.P., Wasielewski, M.R., Forrest, S.R., Ideal diode equation for organic heterojunctions. I. Derivation and application. *Phys. Rev. B*, 82, 15, 155305, 2010.
27. Fezzani, A., Mahammed, I.H., Said, S., MATLAB-based modeling of shading effects in photovoltaic arrays, in: *2014 15th International Conference on Sciences and Techniques of Automatic Control and Computer Engineering (STA)*, IEEE, pp. 781–787, December 2014.
28. Quaschning, V. and Hanitsch, R., Influence of shading on electrical parameters of solar cells, in: *Conference Record of the Twenty Fifth IEEE Photovoltaic Specialists Conference-1996*, IEEE, pp. 1287–1290, May 1996.

Part IV

RF CIRCUIT MODELING

14

HBT High-Frequency Modeling and Integrated Parameter Extraction

Ashish Bhatnagar[1]* **and Rachit Patel**[2]

[1]*Idemia Syscom India Private Limited, Noida, Uttar Pradesh, India*
[2]*Department of Electronics and Communication Engineering, ABES Institute of Technology, Ghaziabad, Uttar Pradesh, India*

Abstract
In this chapter a non-linear, demonstrate for AlGaAs–GaAs HBT's characterization in DC, minimal signal, and noise is mentioned. Conjointly some set of equations square measure thought-about to require into consideration in noise equations. This method is general and can be used to derive parameters from other microwave devices, such as MESFETs and high negatron efficiency semiconductors (HEMTs). This chapter is going to see associate in nursing a method for extracting interconnected parameters that is incontestable for a manufactory HBT from wherever we are going to show some wonderful results. The primary technique for obtaining the model parameter values of analogous circuit models is parameter extraction by fitting the model responses to measurements. Parameter extraction has traditionally relied on large signal measurements and DC parameter. For dc, small signal, and large signal research, the derived versions are appropriate.

Keywords: Parameter estimation, microwave devices, heterojunction bipolar transistors

14.1 HBT High-Frequency Modeling and Integrated Parameter Extraction

Need of greater speed and SNR ratio in analogue of electronic communications has grown and low-noise circuits and high frequency technologies

**Corresponding author*: bhatnagarashish1998@gmail.com

Sangeeta Singh, Rajeev Kumar Arya, B.C. Sahana and Ajay Kumar Vyas (eds.) RF Circuits For 5G Applications: Designing with mmWave Circuitry, (269–278) © 2023 Scrivener Publishing LLC

have become increasingly relevant [1–3]. As a result of this the use of HBT in high speed optical and microwave circuits has sparked a lot of interest. HBTs are increasingly becoming suitable for low-noise electronic equipment around the frequency range 1 GHz to 1000 GHz and to the metric linear unit bands. In particular, in MMIC architecture, accurate system models are crucial to the performance of nonlinear microwave circuits such as amplifiers, oscillators, mixers, receivers, and synthesizers.

14.2 High-Frequency HBT Modeling

The noise association has an impact on the noise parameters in heterojunction bipolar transistors and is thus important for semiconductor modelling. The HBT model's analogous circuit has three parts: intrinsic circuit, system parasitic, and box parasitic. In high-frequency simulation, parasitic devices and packages play a critical role. The model would be unable to correctly forecast ac output if these parasitic were not taken into account. The findings also reveal that the noise parameters were extracted from the segmented transit time for noise, which corresponds well with the tuner-based noise calculation.

TRL adjustment constructs, i.e. 70-220GHz formed on-wafer, are used to de-embed S-parameters. The effects are distinguished at frequencies between 75-110GHz and 140-220GHz. The skill of TRL adjustment in S-parameter processing is demonstrates by achieving good alignment between low and high frequency measurements. There is also a short signal-related circuit for InP DHBT up to 220GHz shown. In the frequency range of 0.2 to 220 GHz, exceptional agreements between observed and calculated data are obtain. In addition, a six-stage amplifier MMIC focused upon on HBT model has seen [4–7]. The amplifier's small-signal gain is larger than 15 dB over the 140-190 GHz frequency range, according to data measured. An electromagnetic simulation-assisted parameter extraction protocol will be address for precise modelling of downscaled transferred-substrate InP HBTs [8–14]. The external parasitic network connecting with transformations and unit electrodes is carefully remove from streamlined three-dimensional EM simulations up to 325 GHz. The outer parasitic network is de-embedded from transistor calculations using an on-wafer multi-line Through-Reflect-Line calibration technique, and the active interface parameters are recovered in a reliable manner. The distributed parasitic network, together with the small-signal model configuration, provides reliable small-signal estimation up to 220 GHz.

14.2.1 DC and Small Signal Models

The outer parasitic network is de-embedded from transistor calculations using an on-wafer multi-line through-Reflect-Line calibration technique, and the active interface variables are recovered in a reliable manner. The distributed parasitic network, together with the small-signal model configuration, provides reliable small-signal estimation up to 220 GHz. Unlike the related articles, we consider using sensitivity analysis of an artificial neural network as the mathematical model [15–18]. This model senses the variables and, as a result, retains its mathematical properties. This technique is applied to a Si/SiGe HBT system. The Figure 14.1 shows the ANN model's design and applicability.

To increase the frequency of unity gain in proportional silicon-germanium heterojunction bipolar transistors, in a commercialized 0.13m SiGe-BiCMOS system network, a f_T-doubler arrangement with three equal SiGe HBTs with similar emitter region of 7 (length) cm^2 is introduced. The IM f_T doubler has a peak f_T of 77 GHz, while an individual IM SiGe HBT has a peak f_T of 53 GHz, demonstrating a 46% improvement in f_T from the f_T-doubler approach. Using Mason's solitary gain, the overall fluctuating frequency of IM f_T is about 158 GHz [1–11]. Furthermore, the small-signal prototype variables of the IM f_T are provided, demonstrating that the IM f_T-doubler architectures could be viewed like an individual semiconductor feature for high-frequency circuit design. For the first time, a SiGe-HBT f_T doubler in IM service inside a commercialize 0.13m BiCMOS technology with a peak f_T of 77 GHz has been demonstrated. A detailed inspection for small-signal prototyping of the IM f_T will be carried out, demonstrating

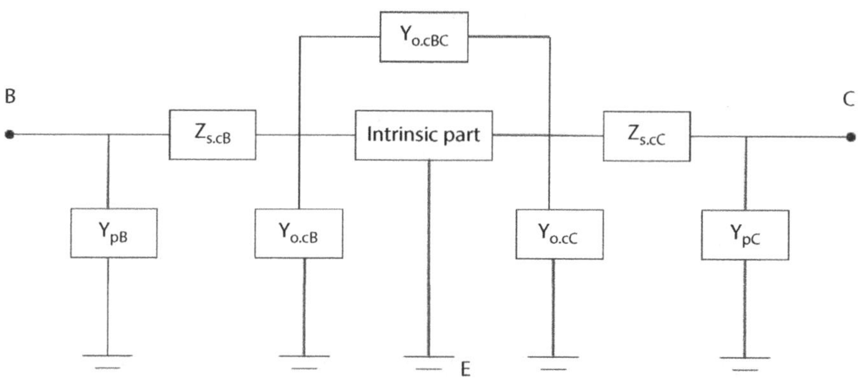

Figure 14.1 In the measuring configuration, the SiGe HBT is represented.

that the f_T-doubler block can be viewed in circuit designs as a single transistor [19, 20]. The f_T-doubler topology's increased current gain, f_T, and speed/bandwidth could be used for range of space-based RF circuits like oscillators and amplifiers.

14.2.2 Linearized T-Model

Resistor between the collector and the emitter to have more precise system output measurement, Rce is applied to the basic T model. For a fair match between modelled and calculated responses, R_{ce} is also crucial in parameter extraction [4, 5]. The fundamental linearized parameters A_0, C_i, R_c, R_e, C_e, and R_{ce} are derived from the dc simulation as functions of V_{be}, V_{ce}, and T_j, are considered in the below example

$$\mathrm{A_0} = \mathrm{A_0}(\varnothing,\mathrm{V_{be}},\mathrm{V_{ce}},\mathrm{Tj}(\varnothing,\mathrm{V_{be}},\mathrm{V_{ce}})) = \frac{\dfrac{\partial I_c'}{\partial V_{be}}}{\dfrac{\partial I_b'}{\partial V_{be}} + \dfrac{\partial I_c'}{\partial V_{be}}}$$

where,

$$I_c' = \mathrm{I_{cf}} - \mathrm{I_{cr}}$$

$$I_b' = \frac{I_{bf}}{BF} - \mathrm{I_{le}}$$

The linearized T model's analogous circuit is shown in Figure 14.2.

A new hybrid-topology-based emitter-length scalable noise model for InP HBT systems and scalable rules for external and internal model parameters are thoroughly developed. The proposed model can use to reliably estimate S parameters and noise efficiency of HBTs of various geometry.

14.2.3 Linearized Hybrid π Model

Figure 14.3 depicts the analogous circuit for the linearized hybrid-model. The internal linearized parameters Y_{be}, Y_{bc}, G_{m0}, and G_0 are properties of the obtained dc simulation model parameters V_{be}, V_{ce}, and T_j like this,

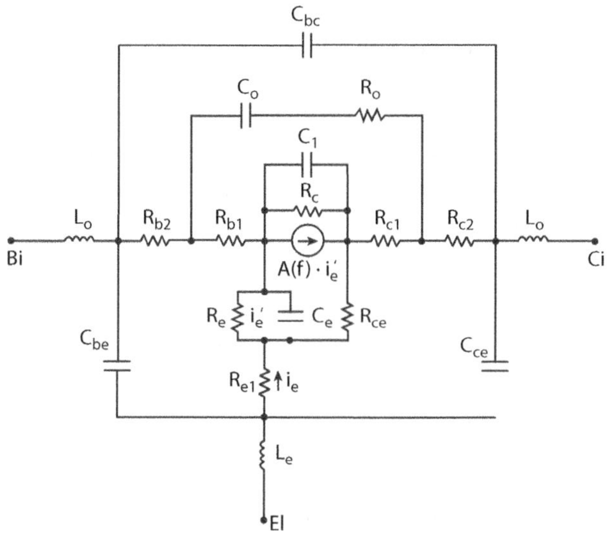

Figure 14.2 The linearized T model using to calculate small signals and noise.

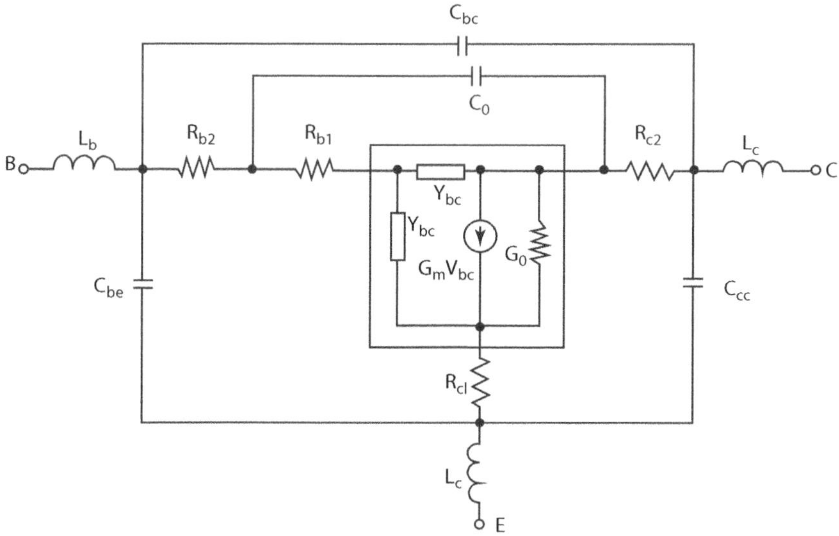

Figure 14.3 Linearized hybrid model is seen for small-signal and noise measurements.

$$G_{m0} = G_{m0}(\varnothing, V_{b_e}, V_{ce}, T_j(\varnothing, V_{be}, V_{ce})) = \frac{dI'_c}{dV_{be}}\bigg|\varnothing, V_{b_e}, V_{ce}, T_j = constant$$

By combining the noise sources' connections as well as their frequency dependence, the hybrid circuit's inherent noise model is established. It is possible to write the normalised noise correlation coefficient.

$$C = \frac{1}{4KT_jB}\begin{bmatrix} \overline{l_b l_b^*} & \overline{l_b l_c^*} \\ \overline{l_c l_b^*} & \overline{l_c l_c^*} \end{bmatrix}$$

Where,

$$\overline{l_b l_b^*} = 4KT_j\mathrm{Re}[Y_{11}]B - 2qI_bB$$

$$\overline{l_c l_b^*} = 2KTj(G_m - 2\mathrm{Re}[Y_{12}])B - 2_qI_cB$$

$$\overline{l_c l_c^*} = 2KTj\mathrm{Re}[Y_{12}]B + 2_qI_cB$$

K-Boltzmann's constant, B-frequency range, q charge of Electron, I_b and I_c inherent base currents and collector currents are respectively. As seen in Figure 14.3, the intrinsic Y-matrix of the intrinsic model, i.e. the region within the strong box, corresponds to Y_{11}, Y_{12}, and Y_{22}, and Re[Y_{ij}] as a result, the actual portion of Y_{ij} is represented. Cross-correlation terms always become zero, meaning that the collector and base currents do not exchange any noise.

Noise parameters which are intrinsic were calculated using matrix C, where Re(C_{12}) and Im(C_{12}) represent the real and imaginary parts of C_{12}, respectively.

$$R_N = C_{11}$$

$$G_{opt} = \sqrt{\frac{C_{22}}{C_{11}} - \left[\frac{Im(C_{12})}{C_{11}}\right]^2}$$

$$B_{opt} = \frac{Im(C_{12})}{C_{11}}$$

$$F_{min} = 1 + 2\sqrt{C_{11}C_{12} - Im(C_{12})^2} + Re(C_{12})$$

14.3 Integrated Parameters Extraction

In this part, we'll look at how to retrieve interconnected parameters by matching the noise calculations and multibias DC S-parameters.

14.3.1 Formulation of Integrated Parameter Extraction

Main function which maximizing is written like this:

$$E_{total}(\varnothing) = E_{dc}(\varnothing) + Es(\varnothing) + E_N(\varnothing)$$

The model parameters to be derived are denoted by Pi, E_{total} is the cumulative error, E_{dc}, E_s, and E_n are respectively DC Error, noise error, S-parameter error and Np is number of ports. N_{dc} is number of dc measurement points.

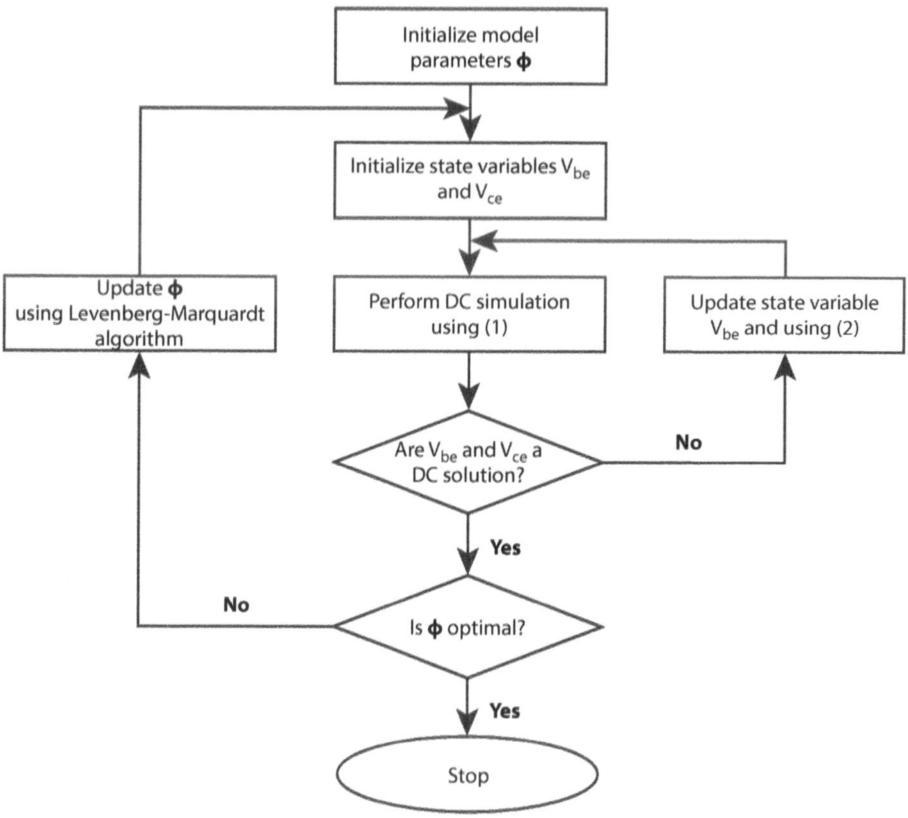

Figure 14.4 Model optimization flow diagram.

N_{Sdc} and N_{sf} number of biasing pints. N_{np} and denotes number of noise parameters, while N_{Ndc} and N_{nf} denote the number of bias points and frequencies where noise measurements are taken [1–7].

$$E_{total}(\emptyset)$$

14.3.2 Optimization of Model

Effective scheme consists of two reduplication cycles. The first iterative method uses DC substitute to determine the DC operating stage, while the second loop adjusts parameter prototypes to adapt sample responses to measurements [10–17]. This approach can be explained using the algorithm below.

Step 1: Set up parameters for model.
Step 2: Set state variables V_{be} and V_{ce} to their default values.
Step 3: Using the Newton iteration, perform DC Simulation for achieving solution for getting State Variables.
Step 4: Calculate the model response using the linearized small-signal circuit.
Step 5: Calculate the optimization goal function.
Step 6: If the modal parameter is optimum, proceed to Step 2; otherwise, change the modal parameter using the Levenberg–Marquardt optimization algorithm.

14.4 Noise Model Validation

The linearized hybrid-pi model is used to validate the noise model, and the linearized T model is used to obtain integrated parameters. For validating model, the ATN-NP5 technique will be used to quantify on-mafer bias- and thermal-dependent noise variable calculations for various foundries [5–12]. To obtain parameters, the above-mentioned system and model were used.

14.5 Parameters Extraction of an HBT Model

The HBT model's parameters are extracted using the linearized T model. Model used at 60 points at DC, 9 biasing points used for S parameter and

Noise, and 17 different frequencies used for extraction of parameters [10–12]. The extraction method is based on a set of close expressions created by not using of any approximation. An equivalent circuit of the forward bias HBT is proposed for the extraction of entry resistances and parasitic inductances. Experiments were carried out on a GaInP/GaAs HBT computer with a 2.25 m emitter, yielding excellent results up to 30 GHz.

Acknowledgement

The authors are thankful to Department of ECE, NIT Patna for providing resources to write this chapter.

References

1. Cai, Q., Gerber, J., Rohde, U.L., Daniel, T., HBT high-frequency modeling and integrated parameter extraction. *IEEE Trans. Microw. Theory Tech.*, 45, 12, 2493–2502, Dec. 1997.
2. Cheng, P. and Shichijo, H., High-frequency noise characterization and modeling of SiGe HBTs. *IEEE Trans. Microw. Theory Tech.*, 66, 12, 5169–5175, Dec. 2018.
3. Lu, H., Cheng, W., Li, O., Kong, Y., Chen, T., Measurement and modeling techniques for InP-based HBT devices to 220GHz. *2018 International Conference on Microwave and Millimeter Wave Technology (ICMMT)*, Chengdu, China, pp. 1–3, 2018.
4. Johansen, T., Doerner, R., Weimann, N., Hossain, M., Krozer, V., Heinrich, W., EM simulation assisted parameter extraction for transferred-substrate InP HBT modeling. *Int. J. Microw. Wirel. Technol.*, 10, 5-6, 700–708, 2018.
5. Venter, J. and Sinha, S., Optimization of 4T pixel structure BiCMOS detectors using y-parameter representation. *2020 IEEE 40th International Conference on Electronics and Nanotechnology (ELNANO)*, Kyiv, Ukraine, pp. 292–296, 2020.
6. Bousnina, S., Mandeville, P., Kouki, A.B., Surridge, R., Ghannouchi, F.M., Direct parameter-extraction method for HBT small-signal model. *IEEE Trans. Microw. Theory Tech.*, 50, 2, 529–536, Feb. 2002.
7. Anholt, R., *Electrical and Thermal Characterization of MESFET, HEMT, and HBT*, Artech House, Norwood, MA, 1995.
8. Vendelin, G., Pavio, A.M., Rohde, U.L., *Microwave Circuit Design Using Linear and Nonlinear Techniȩue*, Wiley, New York, 1990.
9. Kollberg, E.L. (Ed.), *Microwave and Millimeter-Wave Mixer*, IEEE Press, New York, 1984.

10. Rohde, U.L., Whitaker, J., Bucher, T.N.N., *Communication Receiver*, 2nd ed., McGram-Hill, New York, 1997.
11. Rohde, U.L. and Nemkirk, D.P., HBT high-frequency modeling and integrated parameter extraction, in: *RF/Microwave Circuit Design For Wireless Applications*, John Wiley & Sons, Inc., New York, NY, 2000.
12. Fernandez, T., Newport, Y., Zamanillo, J.M., Tazon, A., Mediavilla, A., Extracting a bias-dependent large-signal MESFET model from pulsed I/V measurements. *IEEE Trans. Microw. Theory Tech.*, 44, 372–378, 1996.
13. Anholt, R., Gerber, J., Tayrani, R., Pence, J., A self-heating HBT model for harmonic balance simulators with parameter extraction. *1994 Asia-Pacific Microwave Conf.*, pp. 1029–1032, 1994.
14. Heinemann, B. *et al.*, SiGe HBT with fx/fmax of 505 GHz/720 GHz, in: *IEDM Tech. Dig.*, pp. 3.1.1–3.1.4, Dec. 2016.
15. Johansen, T.K. *et al.*, EM simulation assisted parameter extraction for the modeling of transferred-substrate InP HBTs, in: *Proceedings of the European Microwave Integrated Circuit Conference*, Nuremberg, Germany, October 2017.
16. Lu, H., Zhou, Z., Chengwei, Zhou, J., Chen, T., Chen, C., A new on-wafer multiline thru-reflect-line (TRL) calibration standard design. *APCA (2014), APCAP*, vol. 2, pp. 934–936, July 26-29, 2014.
17. Song, I. *et al.*, The use of inverse-mode SiGe HBTs active gain stages in low-noise amplifiers for the mitigation of single-event transients. *IEEE Trans. Nucl. Sci.*, 64, 1, 359–366, Jan. 2017.
18. Song, I. *et al.*, On the application of inverse-mode SiGe HBTs in RF receivers for the mitigation of single-event transients. *IEEE Trans. Nucl. Sci.*, 64, 5, 1142–1150, May 2017.
19. Devabhaktuni, V., Yagoub, M., Zhang, Q.J., A robust algorithm for automatic development of neural network models for microwave applications. *IEEE Trans. Microw. Theory Tech.*, 49, 2282–2291, 2001.
20. Cen, Y., Zhang, J., Chen, H., Ding, R., Design of a readout circuit incorporating a 12-bit analog-to-digital converter for cooled infrared focal plane array. *Opt. Quant. Electron.*, 51, 392, 1–12, 2019. https://doi.org/10.1007/s11082-019-2084-5.

15

Non-Linear Microwave Circuit Design Using Multi-Harmonic Load-Pull Simulation Technique

Veral Agarwal[1]* and Rachit Patel[2]

[1]Nagarro Software Private Limited, Gurugram, Haryana, India
[2]Department of Electronics and Communication Engineering, ABES Institute of Technology Ghaziabad, Uttar Pradesh, India

Abstract

In this chapter we will learn about the practical applications of the multi harmonic load-pull simulation technique of the design of nonlinear microwave circuits. A structured project using multiharmonic load-pull simulation delve the results of every harmonic ending on the circuit staging also finding flawless load at every harmonic. Systems performance can be improved notably if we use this systematic design procedure. This approach comes up with an efficient means of nonlinear microwave circuit design. Its superiorities are embellishing for the design of frequency doubler and two power amplifiers.

***Keywords*:** Multiharmonic load-pull, non-linear microwave circuit design, frequency doubler, power amplifiers

15.1 Introduction

In recent times as we all know about the speedy evolution of wireless communication. This development makes classification of nonlinear circuits such as power amplifiers, oscillators, and frequency doubler a critical component of the entire communication system. Good designing of such nonlinear circuits can significantly reduce the cost as well as improves the performance of the system.

**Corresponding author*: veral2000@gmail.com

Sangeeta Singh, Rajeev Kumar Arya, B.C. Sahana and Ajay Kumar Vyas (eds.) RF Circuits For 5G Applications: Designing with mmWave Circuitry, (279–288) © 2023 Scrivener Publishing LLC

The harmonic balance (HB) technique has extensively been used for nonlinear microwave circuit design since its execution in commercial Computer Aided Designing (CAD) software also there many other applications have been developed for the HB method like stability analysis [1, 2] and load-pull simulation [3]. Load-pull approach and harmonic load adjusting has used profitably for large-signal device characterization and nonlinear circuit design. Practical usage of the load-pull techniques efficiently is only possible by using systematic scheme. For presenting systematic scheme multiharmonic load-pull simulation using the HB method. Response lineation are imitated by sampling the analogous harmonic impedances are located from the response contours. The process is organized in two crucial steps: 1 Finding the ideal charging at each harmonic and 2 Checking its power elevation of higher harmonics.

15.2 Multi-Harmonic Load-Pull Simulation Using Harmonic Balance

15.2.1 Formulation of Multi-Harmonic Load-Pull Simulation

The multiharmonic load-pull simulation is implemented in a period of the nonlinear simulator, Microwave Harmonica [8], which utilizes systematic HB technique [9]. Circuit topology for the multiharmonic load-pull simulation is often sketched generally as Figure 15.1. where M tuners are positioned near to the M external ports considered. The framework can generally be indicated following:

$$E\,(X, Z) = 0$$

Here E is Vector of HB errors.
X is all harmonic state variable set.
Z is harmonic loads on all external ports set.

$$E_k = A\,(kw_0, Z)\,\phi\,(X, Z) + B\,(kw_0, Z)\,\Psi\,(X, Z) + D\,(kw_0, Z)$$

Where, $0 \leq k \leq NH$ (NH is the digit of modulation utilize in simulation). A & B are model of circuit, D is a pack of driving functions, and Φ and Ψ are receptively, the harmonic vectors of transitory currents I and voltages V at the nonlinear subnetwork ports.

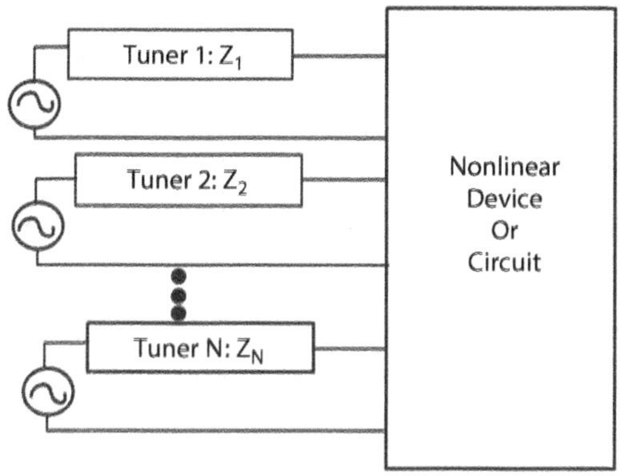

Figure 15.1 Circuit topology for multiharmonic load-pull simulation.

The harmonic loads Z can be recorded in the following matrix from:

$$Z = \begin{bmatrix} Z_1(0w_0) & Z_1(1w_0) & \dots\dots\dots & Z_1\left(N_H w_0\right) \\ Z_1\left(0w_0\right) & Z_1(1w_0) & \dots\dots\dots & Z_1\left(N_H w_0\right) \\ Z_1(0w_0) & Z_1(1w_0) & \dots\dots\dots & Z_1\left(N_H w_0\right) \end{bmatrix}$$

Here, M represent extraneous ports mediated and $Z_i(k\omega 0)$ is the load on k^{th} harmonic and the Ith port with $(0\omega_0)$ representing the dc component.

We use multiharmonic load-pull simulation for the acquisition of the excellent harmonic loading with respect to design specification. An Impedance Sampling technique do utilize in our application of multi-harmonic load-pull simulation.

15.2.2 Systematic Design Procedure

WLOG (Without losing generality), and for elementary demonstration, we will consider the two-port circuit showing in Figure 15.2, where location

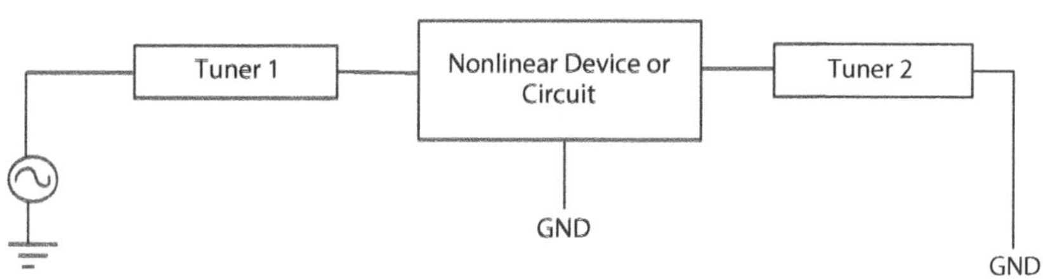

Figure 15.2 Circuit schematic of a typical two port nonlinear circuit.

of tuners is at the Input and load ports. The impedance sampling is acquired through regulating parameters of the tuners, which described as:

$R_i(k\omega 0)$- Resistance at the Ith tuner and the kth harmonic, i = 1, 2, and k = 1, 2, . . .NH.

$X_i(k\omega 0)$- Reactance at the Ith tuner and the kth harmonic, i = 1, 2, and k = 1, 2, . . . NH.

In order to obtain a uniform sampling of points, the tuner impedance is mapped to a reflection coefficient using a reference impedance Zr and the mapping equation.

$$T_i(kw_0) = Z_i(kw_0) - Z_r / Z_i(kw_0) + Z_r$$

The systematic design strategy can be described as follows:

Step 1: Initiation of load-pull paralleling by examining the impedance by central frequency $Z_i(\omega 0)$ and find perfect load $Z_io(\omega 0)$.
Step 2: Fix $Z_i(\omega 0)$ at $Z_io(\omega 0)$ and control output spectrum to witness the chattels of higher harmonic loads on the circuit responses. If the chattels of higher harmonic loads are noteworthy, let k = 2, and go forward towards Step 3. Or then pause.
Step 3: Execute load-pull experiment by sampling the kth harmonic impedance $Z_i(k\omega 0)$ and discover the perfect load $Z_io(k\omega 0)$.
Step 4: Fix $Z_i(k\omega 0)$ at $Z_io(k\omega 0)$ and take care of the output spectrum to examine the chattels of Increased harmonic loads on the responses of circuits. If the chattels of higher harmonic loads are considerable, let k = k + 1, and go forward towards the Step 3. On the other hand stop.

These above steps needed to perform repeatedly for every tuner to attain the most advantageous results. This could be concluded as a load-pull simulation and spectrum analyzing process. The flowchart in Figure 15.3 shows this procedure.

15.3 Application of Multiharmonic Load-Pull Simulation

15.3.1 Narrowband Power Amplifier Design

Here, we are considering two power amplifiers as Amplifier 1 and Amplifier 2. These amplifiers are such designed that they could work at 0.5 GHz. The circuit

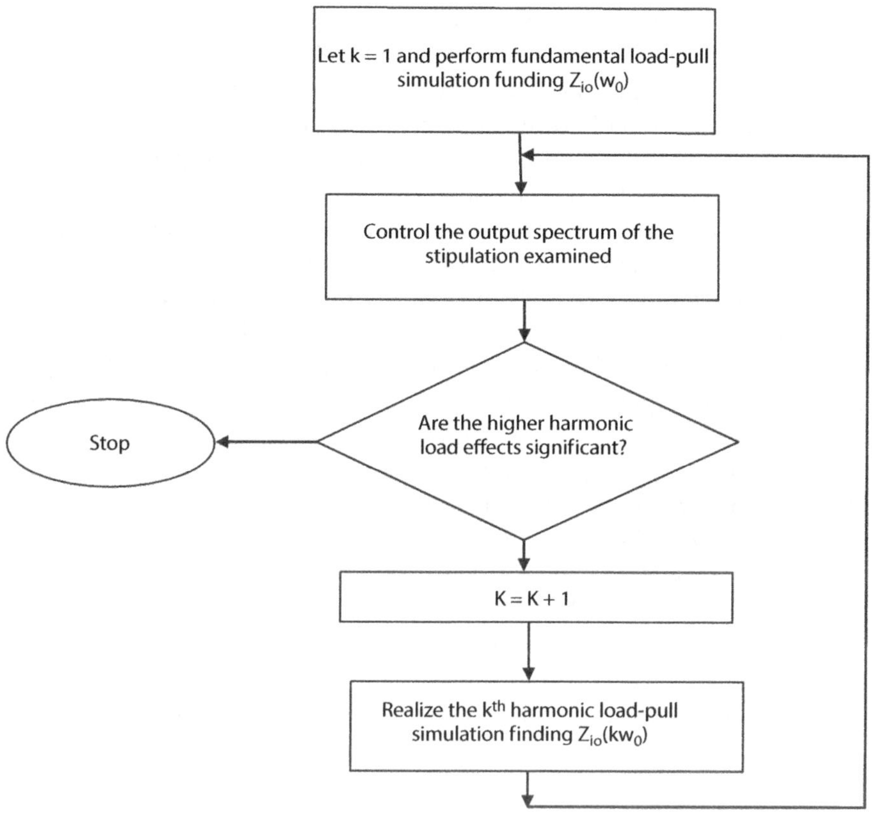

Figure 15.3 Flowchart of design procedure using multiharmonic load-pull simulation.

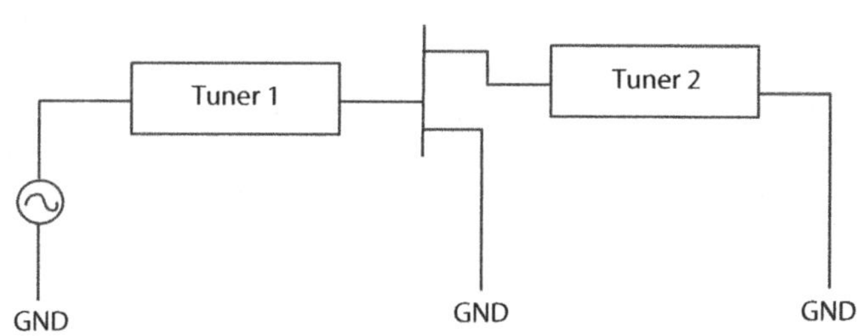

Figure 15.4 Circuit topology for the amplifier design using multiharmonic load-pull simulation.

diagram is shown in Figure 15.4 that is used for amplifier designing. Model Siemens power MESFET CLY10 [10] is used in this prototype. MESFET is patterned by Materka model [11] executed in Microwave Harmonica [9].

In Amplifier 1, MESGET is partisan at Vgs = -1.3 V and Vds = 5 V, amplifier is fabricated like traditional type of class-A with an input power of 10 dBm. Consecutive towards the methodology characterized upon, we perform load-pull experiment by examining $Z_2(\omega 0)$ of Tuner 2 although other harmonic impedances of Tuner 2 are sets permanent at 50 Ω. The load-pull delineation of power gain and power-added efficiency (PAE) are plotted in Figure B-5 [6]. Spectrum of output power at this particular point is plotted in Figure B-6 [6]. By analyzing the output power spectrum of Figure B-6 [6]. Designers found that the power levels at all surpassing harmonics are very small. That is the reason, higher harmonic loads would not have considerably chattels in case of amplifier realization. Designers perform load-pull simulation by $Z_2(2\omega 0)$ of tuner 2 while $Z_2(\omega 0)$ is fixed at $Z_2o(\omega 0)$ for verifying the results. From the designing It is found that, performance with respect to Power Gain and PAE is obtained for purely reactive second-harmonic loads, which is consistent with the results obtained by Berini *et al.* [5]. Output power vs phase of Γ2(2ω0) is shown in Figure B-7 [6], from which we can witness that the inference of $Z_2(2\omega 0)$ is very small and our conclusion is vindicate.

In order for instance the consequences of high harmonic charging on the amplifier realization, MESFET is partisan at Vgs = -2.0 V and Vds = 5 V, and the input power is raised to 20 dB mm in Amplifier 2. The perfect load $Z_{2o}(\omega 0)$ is 4.31+j13.30 for power gain (12.66 dB) and 7.61 + j13.11 for PAE (44.28%). Nevertheless, concession in the middle of the power gain and PAE can be accessed, designers choose 7.61 + j13.11 as $Z_{2o}(w_0)$ and cogitate the PAE as the primary stipulation everywhere in the consecutive procedure. By fixing $Z_2(\omega 0)$ at 7.61 + j13.11 and sampling $Z_2(2\omega 0)$, Designers perform the second-harmonic load-pull examination. The contours of PAE are shown in Figure B-10 [6], which also pinpoints the best for the second-harmonic load will be real reactance near short-circuit point. The effect of the phase of Γ2(2ω0) on PAE is shown in Figure B-11 [6]. The optimal value of $Z_{2o}(2\omega 0)$ is -j6.58, where the phase of Γ2(2ω0) is 195°. The PAE is augmented from 44.28%to 47.18%. The output power spectrum at $Z_2(\omega 0)$ = 7.61 + j13.11 and at $Z_2(2\omega 0)$ = -j6.58 is shown in Figure B-12 [6]. From Figure B-12 [6], we can see that the amplifier realization can be further improvised by a legitimate third-harmonic load. The third-harmonic load-pull simulation is carried out by sampling $Z_2(3\omega 0)$ while $Z_2(\omega 0)$ and $Z_2(2\omega 0)$ are maintained at their optimal values. $Z_{2o}(3\omega 0)$ is found to be -j65.16 at which the value of PAE is 47.66%. The results are as follows showing the procedure we perform (load-pull simulation up to sixth harmonics).

Harmonic load	Optimal value
$Z_2(w_0)$	7.61 + j13.11
$Z_2(2w_0)$	-j6.58
$Z_2(3w_0)$	-j65.16
$Z_2(4w_0)$	-j2.18
$Z_2(5w_0)$	-j28.81
$Z_2(6w_0)$	-j23.32

PAE is 48.13% and power gain is 12.38 dB near the latter architecture. Output Power spectrum of latter architecture is demonstrated in Figures of [6], points that the output power levels at seconds and increased harmonics have been abolished considerably compared to the ones.

15.3.2 Frequency Doubler Design

The MESFET is partisan at pinchoff, with Vgs = -1.9 V and Vds = 5 V. The frequency of input waveform is 5GHz and the input power is 3 dBm. 6 harmonics are considered in the HB Experiment. Both Z_s and Z_l are 50 Ω. The conversion gain (CG) and the spectral purity (SP), are appraised as the design specifications, calculated, respectively, by

$$CG = P_L\,(2w_o) \,/\, P_{avs}(w_o)$$

And

$$SP = P_L(2w_o) \,/\, \Sigma\, P_L\,(kw_o)$$

where PL is the power delivered to the load ZL and Pavs is the available source power. In addition to the load-pull simulation, we also perform source-pull simulation in the design.

Note that the harmonic impedance $Z_i(k\omega 0)$ is the tuner impedance without the load Z_L or the source impedance Z_S, and $\Gamma_i(k\omega 0)$ is the reflection coefficient of the tuner only in this design example. First, the fundamental load-pull simulation is performed by sampling Γ2(ω0) at Tuner 2 while the impedances of all harmonics at Tuner 1 and alternative harmonic impedances of Tuner2 are set permanent at 0 Ω. The delineation of SP with respect to Γ2(ω0), from which we can see that the optimal tuner impedance is infinity, which means that the fundamental harmonic component of the output waveform should be completely reflected by an open circuit. We select 1 ∠ 0 as the optimal point [Γ2o(ω0)], at which the SP is 13.21dB and the CG is 9.75 dB. The second-harmonic load-pull simulation is performed by.

Though a compromise be sampling Γ2(2ω0) with Γ2(ω0) being fixed at Γ2o(ω0). The contours of SP with respect toΓ2(2ω0). The Γ2o(2ω0) is found to be 1 ∠ 50, at which the SP is improved from 13.21 to 18.14 dB and therefore the CG from –9.75 to –5.89 dB. Following an equivalent procedure, the load-pull simulation is performed up to the sixth harmonic. it's found that each one the optimal harmonic impedances of Tuner 2 are infinity, apart from the second harmonic, at which the optimal impedance is found to be j127.23. At the top of the load-pull simulation, the SP is 25.28 dB and therefore the CG is –5.87 dB. The CG cannot be improved considerably by tuning the harmonic impedances of Tuner 2.

The source-pull simulation is administered by sampling the harmonic impedances of Tuner1 using an equivalent procedure as load-pull simulation with all the harmonic impedances of Tuner 2 being fixed at their optimal values. it's found that the foremost dominant harmonic impedance of Tuner 1 is that the fundamental harmonic impedance. The optimal impedances at other harmonics of Tuner 1 end up being 0. The contours of CG with reference to Γ1(ω0) obtained by fundamental source-pull simulation with other harmonic impedances of Tuner1 set to 0 Ω. The Γ1o(ω0) is found to be 1 ∠ 40. The CG is significantly improved from –5.87 to 2.07 dB with a minor 0.5-dB degradation of spectral purity. The input and output voltage waveforms under the optimal termination conditions from which we will see that the frequency of the output voltage is doubled to 10 GHz

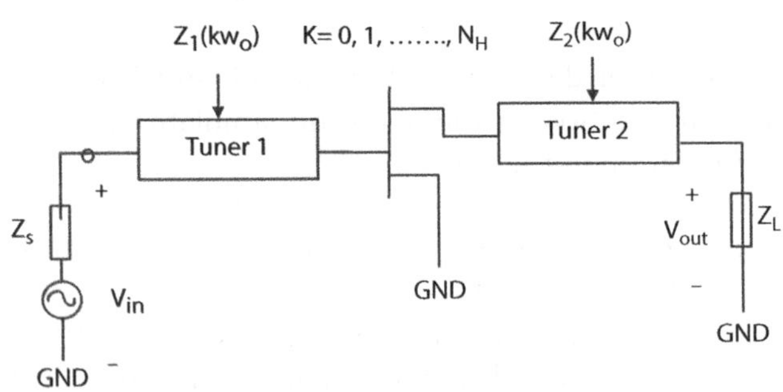

Figure 15.5 Circuit topology for frequency doubler using multiharmonic load-pull simulation.

with an inexpensive gain. The distortion of the input voltage waveform is thanks to the reflection of the higher-order harmonic voltages.

References

1. Rizzoli, V. and Neri, A., State of the art and present trends in nonlinear microwave CAD techniques. *IEEE Trans. Microw. Theory Tech.*, 36, 343–365, 1988.
2. Gerber, J. and Chang, C.R., Application of the harmonic-balance method to the stability analysis of oscillators. *IEEE MTT-S International Microwave Symposium Digest*, pp. 1587–1590, 1997.
3. Cai, Q., Gerber, J., Peng, S., A systematic scheme for power amplifier design using a multi-harmonic load-pull simulation technique. *IEEE MTT-S International Microwave Symposium Digest*, 1998.
4. Pengelly, R. and Saffian, M., Load pull simulation speeds design of wide-band high-efficiency pas [Application Notes]. *IEEE Microw. Mag.*, 17, 42–53, March 2016.
5. Berini, P., Desgagn, M., Ghannouchi, G., Bosisio, R.G., An experimental study of the effects of harmonic loading on microwave MESGET oscillators and amplifiers. *IEEE Trans. Microw. Theory Tech.*, 42, 943–950, 1994.
6. Rohde, U.L. and Rohde, D.P., *R/Microwave Circuit Design for Wireless Applications*, Paterson, New Jersey, John Wiley & Sons, Inc., 2000.
7. Kopp, B. and Heston, D.D., High-efficiency 5-watt power amplifier with harmonic tuning. *IEEE MTT-S International Microwave Symposium Digest*, pp. 839–842, 1988.
8. *Microwave Harmonica Reference Manual*, Ansoft Corporation, Compact Software Division, Elmwood Park, NJ, November/December 1997.
9. Rizzoli, V., Cecchetti, C., Lipparini, A., Mastri, G., General-purpose harmonic balance analysis of nonlinear microwave circuits under multitone excitation. *IEEE Trans. Microw. Theory Tech.*, 36, 1650–1660, 1988.
10. Hajji, R., Beauregard, G., Ghannouchi, G., Multitone power and intermodulation load-pull characterization of microwave transistors suitable for linear SSPAs design. *IEEE Trans. Microw. Theory Tech.*, 45, 1093–1099, 1997.
11. *Einzelhalbleiter Discrete and R Semiconductors, Siemens Data Book 02.97*, Kluwer Technische Boeken, Deventer, Netherlands, 1997.

16

Microwave RF Designing Concepts and Technology

Madhu Raj Kumar[1]* and Neelesh Pratap Singh[1,2]

[1]*Microelectronics and VLSI Design Lab, ECE Department National Institute of Technology Patna, Bihar, India*

[2]*Quality Assurance & Inspection Circle, Bharat Sanchar Nigam Limited, Jabalpur, India*

Abstract

The quality of the passives utilized in many RF designs determines their performance. In silicon, high-quality inductors are unquestionably hard to accomplish. This chapter starts with a review of common RF terminology and concepts, followed by a discussion of technology-related difficulties, with a focus on passives. This chapter is focused on covering the basic microwave RF designing concepts and various existing technologies. The subsequent sections of this chapter cover the various microwave RF figures of merits, its characterization techniques and RF CMOS co-designing approaches.

***Keywords*:** Gain, noise, non-linearity, sensitivity, cut-off frequency, maximum oscillation frequency, input limited frequency

16.1 Introduction

As the performance of any circuit depends on the characteristics of the circuit design hence in this section we have introduced some fundamental terms which characterized any circuit design. The terms which are used for characterizing the circuit are gain, noise, non-linearity and sensitivity of the circuit [1–3].

**Corresponding author*: madhu.ec19@nitp.ac.in

Sangeeta Singh, Rajeev Kumar Arya, B.C. Sahana and Ajay Kumar Vyas (eds.) RF Circuits For 5G Applications: Designing with mmWave Circuitry, (289–312) © 2023 Scrivener Publishing LLC

16.1.1 Gain

In general to measure the gain and signal levels of RF design we use the following terms: The voltage gain, V_{out}/V_{in}, and power gain, P_{out}/P_{in}, are expressed in decibels (dB) the expression used for these term

$$A_{v|dB} = 20log\frac{V_{out}}{V_{in}} \tag{16.1}$$

$$A_{P|dB} = 10log\frac{P_{out}}{P_{in}} \tag{16.2}$$

Both terms are equal (in dB) only when the input and output voltages applied across equal impedances.

16.1.2 Noise

Every system has some noise the performance of RF circuit design is limited by the noise. If a RF systems does not have noise then system (receiver) will detect very small input signal and it allows communication for very long distance but it does not happen because every system have some noise. In this section we have discussed the basic noise which are possibly exist in any system.

i. Shot Noise

In electronic device shot noise generated due to the random fluctuation in the current when the charge carries jumps through the band gap. When charge carries flow and across the band gap they have discrete arrival times these discrete arrival exhibits shot noise. According to Schottky formula the root mean square (rms) value of shot noise current is expressed as

$$i_n = \sqrt{2Iq\Delta B} \tag{16.3}$$

Where, I is the operating DC current, q is the charge of an electron and ΔB is the bandwidth in Hz. The electrons spontaneously exit the cathode and enter the anode, causing shot noise in vacuum tubes. Owing to the presence of a space charge, the tube cannot exhibit shot noise. Resistors and conductor does not exhibit shot noise because in this electrons move due to thermal effect and diffusion.

ii. Flicker Noise

It is associated with signal or process and it is inversely proportional to the frequency (1/f) and falls off as frequency is higher. It exist in all electronics devices and results in different effects [4].

iii. Burst Noise

If sudden transition occurs in two or more discrete current or voltage source more than hundred microvolts randomly and unpredictable then it raise the burst noise. It is also named as popcorn noise due to cracking sounds of it [5].

iv. Transit-Time Noise

If the time taken by electrons to pass from emitter to collector is comparable to the length of the applied signal, it manifests in the transistor. After the frequency at which this effect becomes important increases, the frequency suddenly exceeds all noise sources [6].

16.1.3 Non Linearity

A system is said to be linear if more than one input $x_1(t)$ and $x_2(t)$ are applied to the system then output of the system is expressed as a linear combination of its input

$$ay_1\,(t) + by_2\,(t) = f\,(ax_1\,(t) + bx_2\,(\mathrm{t})) \tag{16.4}$$

Where $ay_1\,(t)$ and $by_2\,(t)$ are the output of the system.

A system is said to be non-linear if its output is as like

$$y\,(t) = a_0 + a_1\,x\,(t) + a_2\,x_2\,(t)\ldots\ldots., \tag{16.5}$$

Where x(t) is input of the system.

The non-linear behavior has to be dealt with at most care while designing the RF front end that can optimize the system.

Effects of Nonlinearity

The system's nonlinearity causes a variety of effects, which we will investigate in this section. Harmonic distortion, gain compression, cross modulation, and intermodulation are the key effects of non-linearity. Although small-signal models can approximate analogue and RF circuits, non-linearity also leads to fascinating and significant phenomena that are not predicted by small-signal models. We look at these phenomena in this section for memoryless systems with order linearity and non-linearity up to order 3[5],

$$y(t) = a_1 x(t) + a_2 x_2(t) + a_3 x_3(t) \tag{16.6}$$

When we applied a sinusoidal signal $x(t) = A\cos(wt)$. The fundamental frequency's amplitude is then expressed as

$$a_1 A + \frac{3a_3}{4} A^3 \tag{16.7}$$

From the above equation it is clear that Because of the third order non-linearity, the small signal gain a_1 varied according to the amplitude of the input.

For $a_3 < 0$, the gain

$$a_1 A + \frac{3a_3}{4} A^3 \tag{16.8}$$

It is decreasing as a function of A. The "1 dB compression rate" (CP_{1dB}) is the input stage the limited signal gain has fallen by one decibel and it can be stated as follows:

$$CP_{1dB} = \sqrt{0.145\left|\frac{a_1}{a_3}\right|} \tag{16.9}$$

It also introduces the higher order the fundamental frequency. Even harmonic terms can be attenuated by differential systems, but it will not remove due to mismatch in symmetry. If we apply two-tone input signal $x(t) = A\cos(w_1 t) + B\cos(w_2 t)$ in this signal first tone is wanted signal and second tone the interfere then the amplitude of the output of the system can be expressed as

$$a_1 A + \frac{3a_3 A^3}{4} + \frac{3a_3 A\ B^2}{2} \tag{16.10}$$

As compared to the interferer, the desired signal is small, A<< B subsequently with $a_3 < 0$ the resulting small signal gain decreases as the interferer's amplitude increases [6]. Consequently, for large value of B the gain of the system drops to zero and this effect is known as blocking. A modulation of the interferer's amplitude is now taken into account, for example $B\,(1 + mcos\,(w_m t))\,cos\,(w_2 t)$ with m < 0, then the output signal's amplitude expressed as

$$a_1 A + \frac{3a_3 A^3}{4} + \frac{3a_3 A\ B^2}{2}\left(1 + \frac{m^2}{2} + \frac{m^2}{2} cos(2w_m\,t) + 2mcos(2w_m\,t)\right) \tag{16.11}$$

Here w_m is the angular frequency of the input signal. The amplitude modulation of the heavy interferer at w_m and $2w_m$ modulates the gain of the desired signal, according to the equation above, and this effect is known as cross modulation.

We've only spoken about the effect of non linearities on the fundamental frequency so far. Non linearities, on the other hand, trigger frequency mixing, which is known as inter modulation (IM) [7]. To discussed this effect we take an input signal having two tone and amplitude of both signal are nearly equal then it is possible to obtain various intermodulation products as

Second order: $a_2\,ABcos\,((2w_1 + w_2)\,t) + a_2\,ABcos\,((2w_1 - w_2)\,t)$

$$\text{Third order: } \frac{3a_3 A^2 B}{4} cos((2w_1 + w_2)t) + \frac{3a_3 A^2 B}{4} cos((2w_1 - w_2)t)$$
$$+ \frac{3a_3 A\ B^2}{4} cos((w_1 + 2w_2)t)$$
$$+ \frac{3a_3 A\ B^2}{4} cos((w_1 - 2w_2)t)$$

Obviously, when $(w_1 = w_2)$ then the third-order intermodulation terms (IM3) reappear in the band of interest and cannot be extracted using a filter, resulting in corrupted output signals. As a measure of this effect, the metric third order intercept point (IP3) has been described. It is known as the point where the fundamental term's amplitude equals the third order inter modulation's amplitude. (Assuming A = B)

$$|a_1|A + \frac{9|a_3|A^3}{4} = \frac{3|a_3|A^3}{4} \tag{16.12}$$

It produces peak voltage rather than power,

$$A_{IP_3} = \sqrt{\frac{4|a_1|}{3|a_3|}} \tag{16.13}$$

for the IP_3, we assume $\alpha \gg \frac{9|a_3|A^2}{4}$ then the output IP_3 is obtained as αA_{IP_3}.

For practical circuits, the assumption $\alpha \gg \frac{9|a_3|A^2}{4}$ holds no longer, and intercept point is always outside of the usable input range. For small input amplitudes, it is calculated by linear extrapolation of observed behavior. The input IP3 (IIP3) can also be established from a single measurements. Applying the IM_3 Products and an identified input power P_{in} we get

$$IIP = \frac{P}{2} + P_{in} \tag{16.14}$$

Where Δ_P the difference between the magnitudes of the fundamentals and the IM3 is the products of the output. The above expression's parameters are all in decibels. The variance between the IIP_3 and the power gain G_p yields the output intercept point OIP_3 i.e. $OIP_3 = \frac{\Delta P}{2} + P_{in}$. From (16.7) and (16.8) the variance between the 1 dB compression point and the input IP3 is nearly equal to -9.6 dB, as can be seen.

When the Equation 16.6 the output of the second non-linear step is cascaded, the result is

$$z(t) = b_1 y(t) + b_2 y^2(t) + b_3 y^3(t) \tag{16.15}$$

Then the overall $A_{IIP3\ is}$ found as below

$$A_{IIP3,all} = \sqrt{\frac{4}{3}\left|\frac{a_1 b_1}{a_3 b_1 + 2a_1 a_2 b_2 a_1^3 b_3}\right|} \tag{16.16}$$

It can be approximated as

$$\frac{1}{A_{IP3,all}^2} = \frac{1}{A_{IP3,1}^2} + \frac{a_1^2}{A_{IP3,2}^2} + \frac{3a_1 a_3}{2b_1} \tag{16.17}$$

Where $A_{IP3,1}^2$ and $A_{IP3,2}^2$ constitute the input for IP_3 points of the first and second stages, respectively.

16.1.4 Sensitivity

The system sensitivity is called the minimum acceptable signal level an instrument can detect by means of an acceptable SNR. The minimum acceptable input power is defined as

$$P_{in,\ min} = B \cdot SNR \cdot P_{noise} \cdot F \tag{16.18}$$

The above expression is also simplified by considering power of the signal matching at the input and under this condition P_{noise} is equal to KT, at room temperature it is equal to the $P_{noise} = -174\ dBm/Hz$, the noise floor at the system's input is equal to B. P_{noise}. The noise floor is the lowest level of noise that a device must be able to detect if it is assumed to be noiseless. The difference of degree of sensitivity to the minimum permissible SNR is found to be the maximum amount of noise that the device should be able to handle. The noise floor is the amount of noise that the device can contribute, or the noise factor, and the total in band noise is the noise level [7].

For a dynamic system noise floor is useful tool to define the dynamic range of the system (DR). The concept of dynamic range in RF circuit design is slightly different, and it is centered on sensitivity and inter modulation, and it is known as spurious - free dynamic range (SFDR) [8, 9].

16.2 Microwave RF Device Technology and Characterization

The best technology for radio frequency IC applications, like integrated circuit technology, is predicted to move in the same direction as digital IC implementations, which is CMOS. The technical requirements for transceiver capabilities, on the other hand, are far more difficult than those for digital integrated circuits. Noise (near carriers), linearity, and gain are performance parameters for RF transceivers, and as a result, integrated circuit technology for RF transceivers is still being developed. CMOS, BiCMOS, GaAs, and bipolar technologies are among the many options available to IC designers [10–12]. Existing technologies have also been improved by inventions such as bipolar and silicon on one (SOA) systems.

16.2.1 Characterization and Modeling

When considering the characteristics of a transistor in terms of some basic and suitable parameters that can be easily calculated, problems often arise. In this section we deal with figure of merit (FOM), it is commonly used term which helps for particular application. However, choice of technology also determined by non-technical factors such as cost, yield, production volume, time to market, etc. In this section we introduce modelling and important technical parameter of RF devices [13–15].

16.2.2 Modeling

It's just as necessary to model a system as it is to characterize it, so it's difficult to analyze and predict transistor-level circuit behavior without them. Even for radio frequencies, bipolar systems have a long background of modeling. As a result, we will not concentrate on bipolar system modeling, but rather on MOS device modeling, which still has this precision. Active study in the area of MOS RF modeling has been going on for the past few years. RF modeling is not possible with "normal" (digital-oriented) MOS models, according to studies. Three key effects are significant at high frequencies: terminal resistance, bulk effect, and non-quasi-static effect [16]. In this section we briefly discussed these effects.

A device's gate resistance is a critical parameter that can be approximated by

$$R_g = \frac{1}{3}\frac{W}{L}R_{poly} \tag{16.19}$$

The sheet resistance of polysilicon, which is employed in the device's gate, is known as R_{poly}. If MOST is properly laid out, the gate resistance can be lowered to a few Ohms. The transconductance and drain current are lowered when resistances are linked in series with the MOST's drain (R_d) and source (R_s). Source resistance lowers noise efficiency while increasing the maximum oscillation frequency. The drain resistance lowers the output impedance of the device. The intrinsic device model requires a proper bulk resistance network since the body impact impacts device functioning. The resistance of the substrate employed beneath in epi-layer determines the values of these resistors [17].

Non-quasi effect occurs when applied signal frequency is same as the maximum operating frequency of MOS device. To demonstrate the non-quasistatic behavior of a MOS device, we apply a sinusoidal signal to its gate terminal. Because of gate modulation, the channel thickness changes uniformly for low frequency input signals. In other words, the channel will have no phase differences. The device's dynamics can be controlled by applying charge equations to static DC equations. For high frequency it is no longer true since channel thickness will be modulated. This effect is strongest for long devices and it can be measurable for those device whose length is nearly equal to the cut-off frequencies. As a result, device equations need to be solved with a non-quasi-static approach [18–21]. Being purely capacitive is one of the important consequences of the non-quasi-static approach, i.e. $Re(Z_i) = 0$. But it is not true for longer at high frequencies because at high frequencies it also has resistive part. In case of saturation region the gate resistance is approximated as

$$R_{in} = \frac{1}{ng_m} \tag{16.20}$$

where $n \approx 5 - 7$. It should be noted that this effect cannot be modelled by including external gate and source resistances.

The resulting model is accurate for RF analysis, but it only includes a subset of non-quasi-static effects. If we want to include complete non-quasi-static operation, you need to split the low frequency MOST into a series of smaller MOSTs in a very special way [1].

16.2.3 Cut-Off Frequency

The cut-off Frequency $f_{T,bipo}$, also known as the transition frequency or unity-current-gain frequency, is probably the most commonly utilized FOM for a bipolar device. The cut-off frequency is $f_{T,bipo}$, which is defined as the common-emitter small signal gain for a short-circuit load dropping to unity. We can utilize the equivalent circuit of a common-emitter circuit to determine the current gain using this approach.

$$r_\pi = \frac{\beta_0}{g_m}, C_\pi = \tau_F g_m + C_{Je}, C_\mu = C_{jc} \tag{16.21}$$

and τ_F is the forward transit time. The junction capacitance of common emitter configuration is C_{je} and C_{jc} are respectively, these are the functions of the voltage V across the junction, which is denoted by the symbol,

$$C_J(v) = \frac{C_J}{\left(1 + \frac{v}{V_J}\right)^{MJ}} \tag{16.22}$$

In the above equation C_J, M_J, and V_J are constants and it is defined by the technology used.

For a common emitter configuration of bipolar device, the small - signal current gain (β) is expressed as

$$\beta(\omega) = \beta_0 \frac{1 + \frac{S}{Z_1}}{1 + \frac{S}{P_1}} \tag{16.23}$$

Here P_1 is pole and Z_1 is zero and it is defined as

$$Z_1 = \frac{-g_{0m}}{C_\mu} \text{ and } P_1 = \frac{1}{r_\pi (C_\pi + C_\mu)} \tag{16.24}$$

β_0 is the static common-emitter gain, it is defined as

$$\beta_0 = g_m r_\pi \tag{16.25}$$

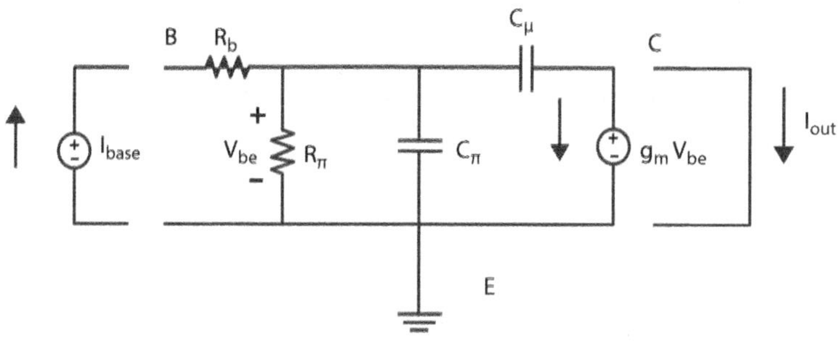

Figure 16.1 Small signal equivalent circuit of a MOSFET with a common emitter configuration and a short circuit load.

For use at high frequencies and in normal operation ($\omega C_{\mu} \ll g_m$) at frequency $f_{T,bipo}$ the current gain β drops to unity and it is defined as

$$f_{T,bipo} = \frac{g_m}{2\pi(C_\mu + C_\pi)} \tag{16.26}$$

Similarly we can also defined unit gain cut-off frequency $f_{T,\,most}$ for MOS

$$f_{T,most} = \frac{g_m}{2\pi(C_{gg} + C_{par})} \tag{16.27}$$

Here we used small signal equivalent circuit as shown in Figure 16.1. It is a combination with the RF extensions. Where, C_{gg} is the intrinsic gate capacitance, C_{gs} is the capacitance of gate source and C_{gd} is capacitance of gate and drain. C_{par} denotes the equivalent capacitances to the substrate, this also involves the overlap and interconnect capacitances. For a MOS device, f_T depends on gate length and for high cut-off frequency it required small gate length.

16.2.4 Maximum Oscillation Frequency

The maximum oscillation frequency f_{max} is defined using the small circuit equivalent of a bipolar device. To find f_{max} we consider the effect of the base resistance, as well as the gate, source, and channel resistances of MOS devices. The maximum oscillation frequency is the one at which the maximum available power gain G_{max} is equal to one. It demonstrates that for maximum oscillation frequency, both voltage and current gain are important. The voltage gain of a bipolar system can be calculated using the small signal equivalent circuit. The input is short-circuited, and the output is filled with current from a current source. The frequency dependent voltage gain A_V is expressed as

$$a_v(\omega) = a_v \frac{1+\frac{S}{Z_1}}{1+\frac{S}{P_1}} \tag{16.28}$$

with

$$Z_1 = \frac{g_m}{C_\mu}, P_1 = \frac{-g_m}{|a_v|C_\mu}, a_v = -g_m R_L \tag{16.29}$$

From the above expression including the current gain expression, $f_{max,bipo}$ can be expressed as

$$f_{max,bipo} = \left(\frac{f_{T,bipo}}{8\pi r_b C_\mu}\right)^{\frac{1}{2}} \tag{16.30}$$

Here, C_μ is the collector - base junction's depletion layer capacitance. In the same way for MOS $f_{max,mos}$ can be expressed as

$$f_{max,mos} = \left(\frac{f_{T,most}}{4\pi C_{Junc}\sqrt{(R_g + R_{in} + R_s)R_{eff}}}\right) \tag{16.31}$$

Where C_{junc} and R_{eff} are drain junction capacitance and the effective bulk resistance of the MOS. From the above expression it is clear that $f_{max,mos}$ is more effected by parasitic elements than $f_{T,most}$ hence device layout is more considered.

16.2.5 Input Limited Frequency

The input limited frequency of the applied voltage signal f_v of the device is related to the frequency dependence transconductance. To derive the relation of transconductance with frequency we use the small signal equivalent circuit. The expression of the transconductance is as below

$$g_m(\omega) = g_m \frac{1+\frac{S}{Z_1}}{1+\frac{S}{P_1}} \tag{16.32}$$

Where

$$Z_1 = \frac{g_m}{C_\mu}, P_1 = -\frac{1}{r_b(C_\mu + C_\pi)} \tag{16.33}$$

The parameters f_v is the cut off frequency of the input voltage signal and it is defined as the frequency at which $g_m(\omega)$ is dropped by 3 dB. For a bipolar device $f_{v,bipo}$ be expressed as

$$f_{v,bipo\approx} = \frac{1}{2\pi r_b(C_\mu + C_\pi)} \tag{16.34}$$

In the same way we can also derived the cut-off frequency for MOS and it can be expressed as

$$f_{v,mos} \approx \frac{1}{2\pi r_g(C_{gg} + C_{par})} \tag{16.35}$$

16.2.6 Output Limited Frequency

When we connect the load resistance R_L to the output and we define the voltage gain as well as output limited frequency. Output limited frequency is the frequency of the signal for which gain drop to 3 dB. It is defined as

$$f_{out,bipo} = \frac{1}{2\pi R_L C_\mu} \tag{16.36}$$

Similarly for MOS output limited frequency is defined as

$$f_{out,mos} = \frac{R_{out} + R_L}{2\pi R_L R_{out} C_{out}} \tag{16.37}$$

Where R_{out} and C_{out} are output resistance and capacitance of the MOS. When we consider first order then Rout approximated as $1/g_{ds}$.

16.2.7 Maximum Available Frequency

For a device it is very important to know the device available bandwidth, f_a. This parameter depends on both input limited frequency and output limited frequency. It is expressed as

$$f_a = \frac{1}{\frac{1}{f_v} + \frac{1}{f_{out}}} \tag{16.38}$$

For a bipolar device, for low current, f_a is determined by f_{out} and high frequency of the device is determined by f_v. $f_{out,bipo}$ is expressed as

$$f_{a,bipo} = \frac{1}{2\pi} \frac{1}{r_b C_\pi + (r_b + R_L) C_\mu} \tag{16.39}$$

16.2.8 Technology Choices

As device modelling meets new hurdles in deep sub-micrometer processes and burgeoning circuit applications, designing analog and radio-frequency (RF) circuits in CMOS technology becomes increasingly challenging. For analog and RF design, the complex collection of characteristics used to describe today's "digital" technology frequently proves inadequate, necessitating numerous more measurements and iterations to arrive at a satisfactory solution. For the RF designing the technological node choice is of great importance. As these nodes defines the CMOS channel length and other device parasitics. Further these parasitic parameters define the circuit operating frequencies.

16.2.9 Double Poly Devices

In this section we will only discussed about double poly devices technology option for a designer when designing a circuit. A polysilicon emitter is used in modern high-performance bipolar devices with a base width of 100 nm or less. The bipolar device's emitter is formed by heavily doping a polysilicon layer, and then the doped polysilicon layer is enabled only as required to obtain repeatable base currents and low emitter resistance. The intrinsic and extrinsic regions of the base region can be separated. The intrinsic base region just beneath the emitter, determines the collector current. Intrinsic base region connects with base through extrinsic base region, and it act as a parasitic component. For decouple the intrinsic and extrinsic base self-aligned process is used. The base is not directly connected to the metal in this technology; instead, contact is made indirectly with the extrinsic base through a p-type polysilicon layer, hence the name double poly method (DPO). On top of a field oxide zone, metal contact to the p-type polysilicon layer is formed. Due to the self-alignment base area obtained is very small, hence a very small extrinsic base-collector junction capacitance and low contact resistance is obtained.

In general, the advantages of this process are low base resistance, reduced contact, very high cut-off frequency and sub-lithographic emitter. Drawback of this technology are, it is more complex, costly and merging with CMOS process is very difficult hence a BiCMOS process based on double poly silicon is very difficult.

16.3 Passive Components

Before designing of an RF IC we need the thorough knowledge of characteristics of passive component used in RF circuit. In general there are three passive components used in any circuit Resistor, Capacitor and Inductor. RFIC design success depends on a detailed understanding of passive components characteristics. In this section we deal with the effect of resistor, Capacitor and Inductor on RF IC design and also its properties since parasitic at large frequency is more important for IC design.

16.3.1 Resistors

Only few resistor options in CMOS technology. The first choice is to use polysilicon interconnect material, which is more resistant than metal. But now a day's most poly is silicided to reduced resistance [22, 23]. Poly is used for moderately small valued resistors because the resistivity drops to about 5-10 ohms per square. The temperature coefficient and its tolerance is poor and nearly equal to 35%, the temperature coefficient (TC) is defined by

$$TC \approx \frac{1}{R}\frac{\partial R}{\partial T} \tag{16.40}$$

Unsilicided poly is also used since its resistivity is very high and its TC can be varies to zero depend on used technology. Source-drain diffusion is also used an option for resistors. MOS transistor also used a variable resistor depends in which region it works typical formula of triode region is

$$r_{ds} \approx \left[\mu C_{ox}\frac{W}{L}[(V_{GS}-V_T)-V_{DS}]\right]^{-1} \tag{16.41}$$

In general, resistance due to contact cells must also be taken into consideration. Typically resistance value depends on the order of half a square.

16.3.2 Capacitors

In VLSI technology, interconnect layers can be used to create traditional parallel plate capacitors. However, regular inter level dielectrics tend to be fairly thick (0.5–1 μm) which reduce the capacitance between layers, results in lower capacitance per unit area 5 x 10^{-5} pF/μm^2. In addition, attention should be paid to the capacitance formed by the bottom plate and the conductors below it (especially the substrate) [24, 25]. This type of parasitic bottom plate capacitance is often 10–30% of the main capacitance, often significantly limiting circuit performance. The standard formula of capacitance is

$$C \approx \varepsilon\frac{A}{H} = \varepsilon\frac{W \cdot L}{H} \tag{16.42}$$

Since it ignores fringes, capacitance is slightly underestimated, but it is correct as long as the plate dimensions are much greater than the plate spacing H. In some cases where this inequality is not well satisfied, a rough first-order correction for the fringing may be provided by adding 2H to each of W and L in computing the area of the plates:

$$C \approx \varepsilon \frac{(W+2H)\cdot(L+2H)}{H} \approx \varepsilon\left[\frac{WL}{H}+2W+2L\right] \tag{16.43}$$

The TC of metal-metal capacitors is very low, usually it is in the range of about 30-50 ppm/degree, and the TC of the oxide permittivity itself dominates because dimensional changes due to temperature are negligible. We can increase the total capacitance per unit area by using multiple interconnect layers. Some process techniques also use five metal layers, so a sandwich structure can be used to quadruple the capacitance. Capacitance can be further increased by utilizing the lateral magnetic flux between adjacent metal wires in a given interconnect layer. This lateral bond is considerable because the allowed adjacent metal spacing has been reduced to a value smaller than the spacing between the layers [2].

A very simple structure of the capacitor having two terminals of the capacitor. The "upper" and "lower" plates constructed from the metal layer alternate to take advantage of the lateral flux. By arranging segments of different metal layers in a complementary pattern, regular vertical flux can also be obtained. The parasitic bottom plate capacitance in lateral flux capacitors is much lower than in a standard parallel plate arrangement, which is a significant feature. This is because it takes up less space for a given total capacitance value. MOS capacitor available in the CMOS process as the gate capacitance is used as a standard alternative. Capacitance of the MOS capacitor is typically depend on oxide thickness and it is in the range of 1-5 $fF/\mu m^2$. Normally MOS capacitor have a small positive TC (at the order of $30ppm/^{\circ}C$), unless the semiconductor is degraded to degeneracy [3]. So gate capacitors have slightly higher TC values. When using a gate capacitor in a CMOS process, it is critical to maintain strong inversion in the transistor; otherwise, the capacitance will be weak, lossy, and highly non-linear. To alleviate this problem, Sometime an accumulation mode MOSFET is built in the IC by using n^+ source-drain diffusions in an n-well. This type of capacitor does not require extra but its characteristics do not controlled by the fabrication. Channel resistance defined by Eqn. 16.40 is used to find the quality factor (Q) and for both types of gate capacitor the Q depend on the channel resistance. Another option is to use a junction capacitance such that it is formed by a p+ region in n-well. Such capacitors are often used to make electronically tuned circuits because the junction capacitance depends on the bias applied. The diodes used in this way are known as varicaps (from variable reactors). The junction capacitance is denoted as C_j and its value is

$$C_j \approx \frac{C_{j0}}{(1 - V_F / \varnothing)^n} \tag{16.44}$$

Where C_{j0} is capacitance at zero bias, V_F is forward bias applied voltage across the junction, Φ is the built -in potential, and n is a parameter depends on the doping profile. For an abruptly doped junction n = ½ and for linearly graded junctions n = 1/3.

Interconnect Capacitance

At radio frequency it is crucial to consider the parasitic capacitance of interconnect. The fringing capacitance is significant and simple formula used to obtain the capacitance formula is unacceptable. We will consider three cases of conductors: i. single wire over a conducting plane of infinite extent; ii. when a single wire in between two infinite plane; and iii. when a wire between two adjacent wires all over a single infinite plane [4–6].

Equation used to obtain the capacitance for case 1 is as:

$$C \approx \varepsilon \left[\frac{W}{H} + \frac{0.15W}{H} + 2.8 \left(\frac{T}{H} \right)^{0.222} 0.15W \right] \tag{16.45}$$

Equation used to obtain the capacitance for case 2 is as:

$$C \approx \varepsilon \left[W \left(\frac{1}{H_1} + \frac{1}{H_2} \right) + 0.77 + 0.891 \left\{ \left[W \left(\frac{1}{H_1} + \frac{1}{H_2} \right) \right]^{0.25} + T^{0.5} \left(\frac{1}{H_1^2} + \frac{1}{H_2^2} \right)^{0.25} \right\} \right] \tag{16.46}$$

Equation used to obtain the capacitance for case 3 is as:

$$C_{total} = C_{single} + 2C_{mutual} \tag{16.47}$$

Where

$$C_{single} \approx \epsilon \left[\frac{1.15W}{H} + 2.8\left(\frac{T}{H}\right)^{0.222} \right] \tag{16.48}$$

and

$$C_{mutual} \approx \left[0.33\frac{W}{H} + 0.83\frac{T}{H} - 0.07\left(\frac{T}{H}\right)^{0.222} \right]\left[\frac{S}{H}\right]^{-1.34} \tag{16.49}$$

16.3.3 Inductors

From an RF circuit perspective, the lack of good inductors is the most prominent drawback of standard IC processes. Active circuits may be able to synthesize the equivalent of inductors, but they are always more noisy, distorted, and consume more power than "real" inductors made by winding wires several times. There are two type of on chip inductor is used Spiral inductors and Bondwire inductors.

Spiral Inductors

Circular spirals are known to provide a slightly higher Q, but geometry other than Manhattan is not supported by many layout tools and is not allowed by many technologies. Most of the spiral is implemented on the topmost metal and the connection to the canter of the spiral is at a lower level. The inductance of such a spiral inductors is a complicated function of geometry and must use the field solver or Greenhouse methods for accurate calculations [7]. The equation used to obtain the inductance of spiral inductors is

$$L \approx \mu_0 n^2 r = 4\pi \times 10^{-7} n^2 \approx 1.2 \times 10^{-6} n^2 r \tag{16.50}$$

Where L is in henneries, n is the number of turns and r is the radius of the spiral in meters.

Further the inductor more improvised in terms of fine tuning an inductor design just prior to verification via a field solver [8]

$$L \approx \frac{37.5\,\mu_0 n^2 a^2}{22r - 14a} \tag{16.51}$$

This equation also applies to a "hollow" spiral inductor, which has one or more of the innermost turns removed to boost Q. Aside from the vast area that could be drained, another significant issue with spiral inductors is the large loss. The skin effect increases DC resistance loss, resulting in RF current distribution in the conductor that is not uniform. As a result, the effective cross-sectional area of the layer is decreased, while its resistance is increased. The complete model of the spiral inductor is described as follows and R_s, δ, C_{ox}, R_1, C_p and C_1 are approximately modelled as

$$R_s \approx \frac{l}{w \cdot \sigma \cdot (1 - e^{-t/\delta})} \tag{16.52}$$

$$\delta = \sqrt{\frac{2}{w\mu_{o\sigma}}} \tag{16.53}$$

$$C_p = n \cdot \omega^2 \cdot \frac{\varepsilon_{ox}}{t_{ox}} \tag{16.54}$$

$$C_{ox} = \omega \cdot l \cdot \frac{\varepsilon_{ox}}{t_{ox}} \tag{16.55}$$

$$R_1 \approx \frac{2}{\omega \cdot l \cdot G_{sub}} \tag{16.56}$$

$$C_1 \approx \frac{\omega \cdot l \cdot C_{sub}}{2} \tag{16.57}$$

Where R_s is series resistance, σ is conductivity of the material, l is the total length, w is width, t is thickness, δ is skin depth, μ_o is permittivity of vacuum, C_p is shunt capacitance, C_{ox} is the capacitance between the spiral and substrate, t_{ox} is the oxide thickness, R_1 is substrate loss, G_{sub} and C_{sub} is fitting parameter and C_1 is the capacitance of the substrate as well as other reactive effects [9].

Bondwire Inductors

In addition to the plane spiral inductor bond wires are also frequently used in the manufacture of inductors. Because standard bond wires are 1 mil (that's 0.001 inches, or about 25 pm) in diameter, Planar spirals have a much smaller surface area per length, resulting in less resistance loss and hence a higher Q value. To reduce parasitic capacitance, it can be placed above any conductive plane and reducing the loss caused by stimulated image currents.

Conclusion

For the 5G communications system the RF system designing is very vital. This chapter demonstrates a panoramic view of basic RF terminology and ideas, then moves on to a discussion of technology-related challenges, with a focus on passives. This chapter covered the fundamentals of microwave RF design as well as existing technology. Finally, the numerous microwave RF figures of merits and RF CMOS co-designing approaches are covered.

Acknowledgement

The authors are thankful to the Microelectronics and VLSI Design Lab, ECE Department National Institute of Technology, Patna, for providing resources to write this chapter.

References

1. Tiemeijer, L.F. *et al.*, MOS model 9 based non-quasi-static small-signal model for RF circuit design. *European Solid-State Device Research Conf. (ESSDERC)*, pp. 652–655, 1999.
2. McCreary, J.L., Matching properties, and voltage and temperature dependence of MOS capacitors. *IEEE J. Solid-State Circuits*, 16, 6, 608–16, 1981.
3. Shahani, A. *et al.*, A 12mW, wide dynamic range CMOS front-end circuit for a portable GPS receiver. *ISSCC Digest of Technical Papers, Slide Supplement*, February 1997.
4. Yuan, C.P. and Trick, T.N., A simple formula for the estimation of the capacitance of two dimensional interconnects in VLSI circuits. *IEEE Electron Device Lett.*, 3, 391–3, 1982.
5. Barke, E., Line-to-ground capacitance for VLSI: A comparison. *IEEE Trans. Comput.-Aided Des. Integr. Circuits Syst.*, 7, 2, 295–298, 1988.
6. Sakurai, T. and Tamaru, K., Simple formulas for two- and three-dimensional capacitances. *IEEE Trans. Electron Devices*, 30, 2, 183–5, 1983.

7. Greenhouse, H.M., Design of planar rectangular microelectronic inductors. *IEEE Trans. Parts Hybrids Packag.*, 10, 2, 101–9, 1974.
8. Wheeler, H.A., Simple inductance formulas for radio coils. *Proc. Inst. Radio Eng.*, 16, 10, 1398, 1928.
9. Gonzales, G., *Microwave Transistor Amplifiers: Analysis and Design*, 2nd edition, Prentice-Hall, Inc., New Jersey, 1997.
10. Tiemeijer, L.F. *et al.*, MOS Model 9 based non-quasi-static small-signal model for RF circuit design. *European Solid-State Device Research Conf. (ESSDERC)*, pp. 652–655, 1999.
11. Wagemans, A. *et al.*, A 3.5mW 2.5GHz diversity receiver and a 1.2mW 3.6 GHz VCO in silicon-on-anything. *IEEE International Solid-State Circuits Conf. (ISSCC)*, pp. 250–251, 1998.
12. Taur, Y. and Ning, T.H., *Fundamentals of Modern VLSI Devices*, Cambridge University Press, New York, 1998.
13. Crols, J., *Full Integration of Wireless Transceiver Systems*, Katholieke Universiteit Leuven, Leuven, 1997.
14. Gonzales, G., *Microwave Transistor Amplifiers: Analysis and Design*, 2nd edition, Prentice-Hall, Inc., New Jersey, 1997.
15. Voinigescu, S.P. *et al.*, Process-and geometry-scalable bipolar transistor and transmission line models for Si and SiGe MMICs in the 5-22GHz range, in: *International Electron Devices and Meeting*, vol. 17, pp. 307–310, 1998.
16. Craninckx, J. and Steyaert, M., A 1.8-GHz low phase noise CMOS VCO using optimized hollow inductors. *IEEE J. Solid-State Circuits*, 32, 5, 736–744, May 1997.
17. Craninckx, J., *Low-Phase-Noise Fully Integrated CMOS Frequency Synthesizers*, Ph.D. Thesis, Katholieke Universiteit Leuven, Leuven, 1997.
18. Amerasekera, A. and Duvvury, C., *ESD in Silicon Integrated Circuits*, John Wiley, Chichester, 1995.
19. Opris, I.E., Bootstrapped pad protection structures. *IEEE J. Solid-State Circuits*, 33, 2, 300–301, Feb. 1998.
20. Ker, M.-D., ESD protection for CMOS output buffer by using modified LVTSCR devices with high trigger current. *IEEE J. Solid State Circuits*, 32, 8, 1292–1296, Aug. 1997.
21. Fujishin, E. *et al.*, Optimized ESD protection circuits for high-speed MOS/VLSI. *IEEE J. Solid-State Circuits*, 20, 4, 594–596, Apr. 1985.
22. Felder, M. and Ganger, J., Analysis of ground-bounce induced substrate noise coupling in a low resistive bulk epitaxial process: Design strategies to minimize noise effects on a mixed-signal chip. *IEEE J. Solid-State Circuits*, 46, 11, 1427–1436, Nov. 1999.
23. Nauta, B. and Hoogzaad, G., Substrate bounce in mixed-mode CMOS IC's. *Workshop on Advances in Analog Circuit Design (AACD)*, 1998.

24. Groeneveld, D.W.J., Groundbounce in CMOS. *Workshop on Advances in Analog Circuit Design (AACD)*, 1993.
25. Leenaerts, D. and de Vreede, P., Influence of substrate noise on RF performance. *European Solid-State Circuits Conf. (ESSCIRC)*, pp. 300–303, 2000.

Index